SpringerBriefs in Physics

SpringerBriefs in Physics are a series of slim high-quality publications encompassing the entire spectrum of physics. Manuscripts for SpringerBriefs in Physics will be evaluated by Springer and by members of the Editorial Board. Proposals and other communication should be sent to your Publishing Editors at Springer.

Featuring compact volumes of 50 to 125 pages (approximately 20,000–45,000 words), Briefs are shorter than a conventional book but longer than a journal article. Thus, Briefs serve as timely, concise tools for students, researchers, and professionals.

Typical texts for publication might include:

- A snapshot review of the current state of a hot or emerging field
- A concise introduction to core concepts that students must understand in order to make independent contributions
- An extended research report giving more details and discussion than is possible in a conventional journal article
- A manual describing underlying principles and best practices for an experimental technique
- An essay exploring new ideas within physics, related philosophical issues, or broader topics such as science and society

Briefs allow authors to present their ideas and readers to absorb them with minimal time investment. Briefs will be published as part of Springer's eBook collection, with millions of users worldwide. In addition, they will be available, just like other books, for individual print and electronic purchase. Briefs are characterized by fast, global electronic dissemination, straightforward publishing agreements, easy-to-use manuscript preparation and formatting guidelines, and expedited production schedules. We aim for publication 8–12 weeks after acceptance.

A. K. Kapoor · Prasanta K. Panigrahi ·
S. Sree Ranjani

Quantum Hamilton-Jacobi Formalism

 Springer

A. K. Kapoor
Chennai Mathematical Institute
Chennai, Tamil Nadu, India

S. Sree Ranjani
STM Journals
Elsevier Publishing Company
Chennai, Tamil Nadu, India

Prasanta K. Panigrahi
Department of Physical Sciences
IISER Kolkata
Mohanpur, West Bengal, India

ISSN 2191-5423   ISSN 2191-5431 (electronic)
SpringerBriefs in Physics
ISBN 978-3-031-10623-1   ISBN 978-3-031-10624-8 (eBook)
https://doi.org/10.1007/978-3-031-10624-8

This Springer imprint is published by the registered company Springer Nature Switzerland AG
The registered company address is: Gewerbestrasse 11, 6330 Cham, Switzerland

This work is dedicated to
Prof. V. Srinivasan
for numerous enlightening sessions.

Preface

The quantum Hamilton-Jacobi formalism was proposed as an attempt to have a quantum version of canonical transformations. As a part of this proposal Leacock and Padgett used the quantum version of the classical action angle variable approach for the bound states on a firm basis. This proposal has been extensively studied and is now a fully developed framework for solution of one-dimensional problems.

The main aim of this monograph is to make this formalism available to a wider audience having a working knowledge of quantum mechanics and of complex variables. Keeping this in mind, sufficient details of, for example, the definition and usage of functions with branch points, residue at infinity, etc. have been given.

This formalism also makes use of a few results from theory of linear differential equations and properties of solutions of Riccati equations. In order to facilitate an understanding of working of the formalism, clear and precise statements of mathematical results and references, where further details can be found, have been included.

It is recommended that Chap. 2 is studied and understood first. The remaining chapters are independent and can be read in any order.

In Chap. 3 we give a few selected examples of solving for eigenvalues and eigenfunctions for the Morse oscillator, radial oscillator and particle in a box. Several other potential problems have been investigated, and the solutions are available in references cited at the end of Chap. 3. Numerous investigations have been carried out on the exactness of supersymmetric Wentzel-Kramers-Brillouin approximation. The quantum Hamilton-Jacobi formalism gives a simple understanding of this result in terms of the analytic structure of the solutions in the extended complex plane. This is explained in the last section of Chap. 3.

Some exotic potential models are included in Chap. 4. The solutions of these and other related models are available only in research papers, or in advanced monographs. These solutions require, as a prerequisite, an understanding of topics beyond the scope of undergraduate textbooks. But, the treatment of these models, within the quantum Hamilton-Jacobi formalism as given in Chap. 4, does not require any more technicalities other than those already given in Chap. 2 showing the simplicity and power of the formalism.

In Chap. 5 some aspects of supersymmetry in quantum mechanics, shape invariance of potentials and isospectral deformation are presented from the point of view of this formalism. This relationship, in turn, is used to provide a simple, transparent and elegant framework for construction of quantum mechanical models related to the exceptional orthogonal polynomials discovered and intensely studied in the previous decade.

The $\mathcal{PT}$ symmetric complex potentials have been a subject of intense investigations and have generated a lot of interest. The quantum Hamilton-Jacobi formalism provides a natural framework to study various properties of $\mathcal{PT}$ symmetric models. A detailed study of one such model is taken up in the Chap. 6, Sect. 6.2. In Sect. 6.3 the role of $\mathcal{PT}$ symmetric models for optical systems is briefly explained.

The applications of the formalism are not restricted to the solution of quantum mechanical models. Several specialized and interesting research investigations related to this formalism have been carried out. For reasons of constraint on the scope and the size, it has not been possible to touch upon those research studies.

In order to show that the quantum Hamilton-Jacobi formalism is not about solving one-dimensional models only, in Chap. 7 we have given a summary of the extensive results on classification of exactly solvable and quasi-exactly models by Ushveridze. An investigation of other aspects of higher dimensional models offers many challenging problems and is still open.

Chennai, India   A. K. Kapoor
Mohanpur, India   Prasanta K. Panigrahi
Chennai, India   S. Sree Ranjani
May 2022

Acknowledgements

This book is a culmination of over two decades of research done with our various collaborators R. S. Bhalla, K. G. Geojo, S. Jayanthi and R. Sandhya. In these collaborations, we have obtained interesting and exciting results. We have tried to include as many of these results here but due to lack of space we could only give a passing reference to some. But, all the related references have been included in the bibliography. We thank them for this long and fruitful association.

A. K. Kapoor thanks Pankaj Sharan and E. Harikumar for painstakingly going through large parts of the manuscript and giving their valuable feedback. This led to substantial improvement in the presentation of this monograph. I wish to acknowledge continuous support and encouragement received from A. P. Balachandran and H. S. Mani.

Prasanta K. Panigrahi acknowledges Subimal Deb, Abhinash Roy and Varun Srivastava for enlightening discussions and their help.

S. Sree Ranjani acknowledges financial support from the Science and Technology Research Board (SERB), Government of India under the extramural project scheme, Project Number: EMR/2016/005002. The two decades of research which is now part of this book would not have been possible without the unconditional love and support from Ravi. I am indebted to Srivalli and Lakshmi Prasanna (Madhu), for standing by me through all the highs and lows and for being my pillars of strength. Valli thank you for being my anchor throughout, and Madhu thank you for always putting things in perspective and stopping me from unravelling. I thank Shreecharan for being a great work buddy, and I appreciate all the useful discussions related to work and otherwise over hundreds of cups of chai. I thank my family for their patience and understanding. Finally, I want to thank my mom Jayashree and my daughter Nirjhari, for making this whole journey meaningful and worthwhile.

We thank B. Ananthanarayan, for inviting us to write for Springer Briefs and for guiding us throughout the whole process. We thank Lisa Scalone for coordinating at different stages of manuscript preparation and for promptly replying to all our queries. We thank Ashok Arumairaj and his team for the technical support.

Contents

Acronyms

COPs   Classical orthogonal polynomials
CT   Canonical transformation
EOPs   Exceptional orthogonal polynomials
ES   Exactly solvable
ISD   Isospectral shift deformation
LPBC   Leacock-Padgett boundary condition
LPEQC   Leacock-Padgett exact quantization condition
PT   Parity and time reversal (symmetry)
QES   Quasi-exactly solvable
QHJ   Quantum Hamilton-Jacobi
QMF   Quantum momentum function
SI   Shape invariance
SUSY   Supersymmetry
SUSYQM   Supersymmetric quantum mechanics
SWKB   Supersymmetric Wentzel-Kramers-Brillouin (approximation)
TE Mode   Transverse electric mode
TM Mode   Transverse magnetic mode

Chapter 1
Quantum Hamilton-Jacobi Formalism

1.1 Introduction

The quantum Hamilton-Jacobi (QHJ) formalism is one of the independent formalisms of quantum mechanics [1], whose roots lie in the esoteric formulation of classical mechanics, namely the Hamilton-Jacobi theory [2, 3]. The latter is intimately connected to the canonical transformation theory, whose development led to the action-angle formulation of classical mechanics. This form of classical dynamics played an important role in the development of quantum mechanics through the Bohr-Sommerfeld quantization rule and its relativistic version. This area known as semi-classical quantization, is full of many important developments and has been particularly succinct in describing the bound states, arising from the periodic motion captured by the angle variables that keep the action constant. In particular, works of Synge [4, 5] and of Einstein, Brillouin and Keller [6–8] have been important landmarks. The area continues to be of active research interest even after several decades [9, 10] have passed since the birth of quantum mechanics.

The discovery of wave mechanics by Schrödinger, illustrated the fundamental importance of the Hamilton-Jacobi formulation of classical mechanics. The action characterizes a class of trajectories in contrast to the conventional coordinates in Newton's equation representing one of these. The Hamilton-Jacobi formalism was christened as the "golden road to quantization".

The formal development of the quantum Hamilton-Jacobi theory started from the canonical transformation theory of Jordan and Dirac [11–13]. While both Jordan and Dirac arrived at the canonical transformation equations, which are in the same form as the classical equations, it was Dirac who introduced the quantum action analogous to the classical action function. Later on, Feynman showed that a path integral approach to quantum mechanics can be formulated starting with the action as the key ingredient [14]. Schwinger also made fundamental use of the quantum action-angle variables [15].

It was Leacock and Padgett who attempted to develop a quantum canonical theory on the lines of the classical Hamilton-Jacobi theory in their 1983 papers [16] and

© The Author(s), under exclusive license to Springer Nature Switzerland AG 2022
A. K. Kapoor et al., *Quantum Hamilton-Jacobi Formalism*,
SpringerBriefs in Physics, https://doi.org/10.1007/978-3-031-10624-8_1

[17]. This formalism, called the QHJ formalism, has a formal correspondence with its classical counterpart in the $\hbar \to 0$ limit. Here the quantum action and the quantum momentum function (QMF), defined analogous to their classical counterparts, play a central role.

As in the classical case, the action angle formalism has been developed to study the bound states. For these systems, the quantum action is equal to an integral multiple of Planck's constant, giving an exact quantization condition. The equation satisfied by the QMF is known as the QHJ equation. This is a nonlinear equation, of the form of the Riccati equation, and can be transformed into the Schrödinger equation using the Cole-Hopf transformation. The exact quantization condition with inputs from the QHJ equation allows us to calculate the energy levels of a given system, without having to solve the equation of motion. Here the correspondence between the classical momentum function and the QMF, in the limit $\hbar \to 0$, acts as a crucial boundary condition required to arrive at the results.

While not much progress has been made in the program of quantum canonical transformations, the formalism of Leacock and Padgett is now a well-developed formalism for one dimensional problems in quantum mechanics.

This monograph takes the reader through various stages of development culminating in its present final form thereby increasing the scope of its applications.

Shortly after the papers by Leacock and Padgett, the QHJ formalism was applied to various exactly solvable (ES) models and the energy eigenvalues were obtained [18]. The potential models investigated were harmonic oscillator, Morse oscillator, Scarf potentials, etc., to name a few.

The original method, as prescribed by Leacock and Padgett, enabled one to calculate only eigenvalues without solving the equation of motion. Subsequently, a simple and elegant method was evolved from the QHJ formalism, which also allows one to calculate the eigenfunctions in a straightforward fashion.

In addition to ES models, this method has been used to investigate many types of potential models like the quasi-exactly solvable (QES) models, periodic potential models and potentials symmetric under combined parity and time reversal ($\mathcal{PT}$ symmetric models for short) [19–24].

During our investigations, several simpler alternatives to the original boundary and the exact quantization conditions have been found. Thus making the formalism simple enough to be included in the first-level quantum mechanics courses. Interesting and surprising results were found, for example those regarding the existence and the behaviour of (complex) zeros of the eigenfunctions of QES and periodic QES models. These results were in contrast with the behaviour expected from oscillation theorems.

It should be noted that all the models discussed in this book are either one dimensional models or separable models in higher dimensions. The fact that this formalism works in complex plane turns out to be a big advantage. For all the models studied, the QMF turns out to be a meromorphic function, defined on the extended complex plane, with a finite number of poles without exception. These facts coupled with a few well-known theorems in complex variable theory allow us to obtain the required results even for potentials which appear intractable. The QHJ formalism also allows us to construct new ES rational potentials and obtain their solutions in terms of the exceptional orthogonal polynomials.

In the next section we present an outline of QHJ formalism, indicating the main points of the theory.

1.2 Quantum Hamilton-Jacobi Formalism

The QHJ formalism to be discussed is about solving the Schrödinger equation for a particle moving in a one-dimensional potential well $V(x)$

$$-\frac{\hbar^2}{2m}\frac{d^2\psi(x)}{dx^2} + V(x)\psi(x) = E\psi(x). \tag{1.1}$$

The first step in setting up the QHJ scheme is to switch over from the Schrödinger equation to QHJ equation. For this purpose quantum action, $S(x, E)$, is introduced by

$$\psi = e^{iS(x,E)/\hbar}, \qquad S(x, E) = \frac{\hbar}{i}\log\psi. \tag{1.2}$$

The quantum action $S(x, E)$ obeys the equation

$$\frac{1}{2m}\left(\frac{dS(x, E)}{dx}\right)^2 - \frac{i\hbar}{2m}\frac{d^2S(x, E)}{dx^2} + V(x) - E = 0. \tag{1.3}$$

This equation is known as the quantum Hamilton-Jacobi (QHJ) equation. In the limit $\hbar \to 0$, it reduces to the classical Hamilton-Jacobi equation with the classical action replacing the quantum action function.

The central quantity of interest, called the quantum momentum function (QMF), $p(x, E)$, is defined as

$$p(x, E) = \frac{dS_E}{dx} = -i\hbar\left(\frac{1}{\psi}\frac{d\psi}{dx}\right), \tag{1.4}$$

which obeys the equation

$$\frac{p^2(x, E)}{2m} - \frac{i\hbar}{2m}\frac{dp(x, E)}{dx} + V(x) - E = 0. \tag{1.5}$$

1.2.1   Working in the Complex Plane

The QHJ formalism takes a big break from standard quantum mechanics by taking x to be a complex variable. The entire formalism makes an essential use of theory of functions of a complex variable. We look for solution of Eq. (1.5) in the complex domain. The QHJ formalism works independently of the Schrödinger wave mechanics.

Two new ingredients, viz. boundary condition on QMF and an exact quantization condition, are taken up in the next two subsections. These replace the standard requirements on the solution of the Schrödinger equation.

1.2.2   Boundary Condition

The QHJ equation for QMF is a quadratic equation and therefore leads to more than one solution. Thus there is a need to formulate a boundary condition to pick the correct solution. Leacock and Padgett use the fact that in the limit $\hbar \to 0$, the QMF should match with the classical momentum function $p_{cl} = \sqrt{2m(E - V(x))}$.

It is to be noted that appearance of square root in the classical momentum function p_{cl} makes it a multiple valued function of complex variable x. Its complete definition requires that a branch cut and a branch must be specified. Leacock and Padgett proposed that the branch cut for the classical momentum function must be chosen to lie in the classically accessible region, and its value must be positive just below the cut in the complex x plane. We will call this requirement as the Leacock-Padgett boundary condition (LPBC), to distinguish it from other ways of selecting the residues in later chapters,

$$\lim_{\epsilon \to 0} \ p_{cl}(x - i\epsilon) > 0 \text{ for } x \text{ such that } (E - V(x)) > 0. \qquad (1.6)$$

In view of rather technical nature of multiple valued functions, we will give a detailed explanation of the definition of the classical momentum function for harmonic oscillator in the complex plane in Sect.2.5.2 and in Sect.2.5.3. The use of LPBC will appear again in Sect.4.3.5 For many applications to be discussed later, we shall use suitable requirements, other than LPBC, to select the physically acceptable solutions.

1.2.3   Exact Quantization Condition

Once a solution for QMF has been found, the energy eigenvalues are determined by making use of an exact quantization condition. The exact quantization condition proposed by Leacock and Padgett is

Fig. 1.1 Contour for action integral $J(E)$

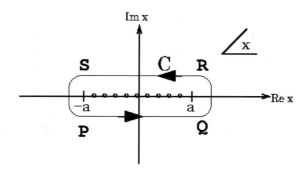

$$J(E) \equiv \frac{1}{(2\pi)} \oint_C p(x, E)dx = n\hbar, \qquad n = 0, 1, 2, \cdots, \qquad (1.7)$$

where C is a contour, see Fig. 1.1, in the complex x- plane enclosing the nodes of the wave function, with n denoting the number of nodes of the latter. The integral $J(E)$ will be called the action integral.

It is useful to mention here that the quantized eigenvalues can be determined in many other ways also. The quantization condition Eq. (1.7) will be called Leacock-Padgett exact quantization condition (LPEQC) to distinguish it from other equivalent requirements that appear in the later chapters of the book. Details of the applications of above two conditions to compute eigenvalues are reserved for Sects. 2.5 and 4.3.

1.3 Plan of the Book

The book is divided into seven chapters. In the second chapter, useful mathematical results from the complex variable theory and from the theory of differential equations are explained. We discuss the simplest example, the harmonic oscillator in Sect. 2.5 and show how we can obtain the eigenvalues following the original scheme given by Leacock and Padgett [16, 17]. This is followed by an alternate, simpler method to solve the harmonic oscillator problem in Sect. 2.6 bypassing the use of LPBC and LPEQC. We also show how eigenfunctions can be obtained from QHJ equation once eigenvalues are known.

In the third chapter, we concentrate on ES models and show how we can go beyond the Leacock-Padgett prescription and obtain the solutions. By doing this we show that both the eigenvalues along with eigenfunctions can be obtained. The models discussed are the Morse potential, the radial oscillator and particle in a box. In the last section of this chapter, we illustrate how LPEQC in the QHJ formalism leads to an understanding of the exactness of the well-known supsersymmetric WKB approximation. This involved the use of the analytic structure of the QMF in the complex plane.

In the fourth chapter we look at potential models which are not amenable to the QHJ formalism in its original form. These models include a potential exhibiting both band spectrum and bound state spectrum, a potential with two phases of SUSY, a (QES) model and finally the periodic Lamé followed by the associated Lamé potentials. Each of these potentials posed a different challenge and we show how each of these could be surmounted and the solutions obtained.

In the fifth chapter, we discuss the construction of new ES rational potentials. We present a method, namely the isospectral shift deformation, to construct rational extensions of the ES models. We have explained the method by constructing the rational extension of the radial oscillator potential and obtained the corresponding solutions in terms of L1 type exceptional Laguerre polynomials (EOPs).

Chapter 6 concerns with models with parity and time reversal symmetry. In these models the potential is complex and the eigenvalues may be real or complex. In Sect. 6.2 the QHJ formalism is applied to a case of $\mathcal{PT}$ symmetric complex potential. The structure of poles of the QMF throws significant light on the nature of the eigenvalues. These are real for the unbroken phase of $\mathcal{PT}$ symmetry and are complex for the broken phase. A study of $\mathcal{PT}$ symmetry has been at the center stage of many developments in optics. We explain this connection in the in the rest of the chapter.

All applications discussed by us concern only analytical solutions. Numerical studies of bound states as well as scattering states for potential models are not included for lack of space and scope of the book. Numerical schemes for bound state and scattering problems have been discussed in [25–27] for several models.

An alert reader would have noticed that all the applications, discussed in this work, are to the exact solutions to one-dimensional models only. The last chapter of the book concerns classification of exactly solvable and quasi-exactly solvable models in one and higher dimensions. The basic equation in this work by Ushveridze, also in the complex domain, is seen to be an extension of the QHJ equation to an equation with several spectral parameters [28]. In our language, the central assumption of Ushverdze's work translates into the QMF being a rational function, a property that the reader will come across repeatedly in all the applications throughout this monograph. Once the form of the QMF is fixed, a requirement the wave function must belong to a physical Hilbert space gives a complete classification of the ES and QES models after a lengthy analysis. This work has been summarized in the last chapter. In the concluding section of Chap. 7 and of the book, we list a set of challenging problems that remain open and require further investigation.

References

1. Daniel, S.F., et al.: Nine formulations of quantum mechanics. Am. J. Phys. **70**(3), 288–297 (2002)
2. Goldstein, H.: Classical Mechanics. Pearson Education India (2002)
3. Rana, N.C., Joag, P.S.: Classical Mechanics. Tata McGraw-Hill (1991)
4. Synge, J.L.: Phys. Rev. **89**, 467 (1953)
5. Synge, J.L.: In Handbuch der Physik, vol. III/1. In: Flugge, S. (ed.). Springer, Berlin (1960)

6. Einstein: Phys. Ges. Verh. XIX, 82 (1917)
7. Brillouin, L: J. Phys. Radium **7**, 353 A926)
8. Keller, J.B. : Ann. Phys. (N.Y.) **4**, 180 A958)
9. Berry, M.V., Mount, K.E.: Semiclassical approximations in wave mechanics. Rep. Prog. Phys. **35**, 315 (1972)
10. Brack, M., Bhaduri, R.K.: Semiclassical Physics. Addison Wesley Publishing Co., Inc Massachusetts (1997)
11. Jordan, P.: Über Kanonische Transformationen in der Quantenmechanik. Zeitschrift für Physik **37**(4), 383–386 (1926)
12. Dirac P.A.M.: The Principles of Quantum Mechanics, Number 27. Oxford University Press (1981)
13. Utiyama, R.: On the canonical transformation in quantum theory. Prog. Theor. Phys. **2**(3), 117–126 (1947)
14. Feynman, R.P. et al.: Quantum Mechanics and Path Integrals. Dover Publications Inc. (2010)
15. Schwinger, J.: Quantum variables and the action principle. Proc. Nat. Acad. Sci. U.S.A. **47**(7), 1075 (1961)
16. Leacock, R.A., Padgett, M.J.: Hamilton-Jacobi action angle quantum mechanics. Phys. Rev. D **28**(10), 2491 (1983)
17. Leacock, R.A., Padgett, M.J.: Hamilton-Jacobi theory and the quantum action variable. Phys. Rev. Lett. **50**(1), 3 (1983)
18. Bhalla, R.S., et al.: Quantum Hamilton-Jacobi formalism and the bound state spectra. Am. J. Phys. **65**(12), 1187–1194 (1997)
19. Bhalla, R.S., et al.: Quantum Hamilton-Jacobi analysis of phases of supersymmetry in quantum mechanics. Int. J. Mod. Phys. A **12**(10), 1875–1894 (1997)
20. Bhalla, R.S., et al.: Energy eigenvalues for a class of one-dimensional potentials via quantum Hamilton-Jacobi formalism. Mod. Phys. Lett. A **12**(05), 295–306 (1997)
21. Bhalla, R.S., et al.: Exactness of the supersymmetric WKB approximation scheme. Phys. Rev. A **54**(1), 951 (1996)
22. Sree Ranjani, S. et al.: Bound state wave functions through the quantum Hamilton-Jacobi formalism. Mod. Phys. Lett. A. **19**(19), 1457–1468 (2004)
23. Cooper, F., et al.: Supersymmetry and quantum mechanics. Phys. Rep. **251**(5–6), 267–385 (1995)
24. Cooper, F., et al.: Supersymmetry in Quantum Mechanics. World Scientific Publishing Company (2001)
25. Leacock, R.A.: Action variable perturbation theory. Phys Lett. **104A**, 184–188 (1984)
26. Chou, C.C., Wyatt, R.E.: Computational method for the quantum Hamilton-Jacobi equation: bound states in one dimension. J. Chem. Phys. **125**, 174103–10 (2006)
27. Chou, C.C., Wyatt, R.E.: Computational method for the quantum Hamilton-Jacobi equation. One Dimen. Scattering Probl. Phys. Rev. E **74**, 066702–9 (2006)
28. Ushveridze, A.G.: Quasi-Exactly Solvable Models in Quantum Mechanics. Institute of Physics Publishing, Bristol and Philadelphia (1993)

Chapter 2
Mathematical Preliminaries

2.1 Introduction

In the first two sections of this chapter we explain results from theory of functions of a complex variables and from theory of differential equations. These results will be needed for different applications in later chapters.

In the last two sections we present a solution of the harmonic oscillator problem using two different methods. In the first method, Sect. 2.5, the original proposal of Leacock and Padgett is followed using LPBC and LPEQC. For use of this approach for several other potentials in we refer the reader to [1, 2]. It will become necessary to use this approach for a model exhibiting two phases of supersymmetry, see Sect. 4.3. Simpler alternatives to LPBC and LPEQC will be used in other sections.

We have attempted to give full mathematical details for the harmonic oscillator problem; the reader may appreciate the original method as proposed by Leacock and Padgett [3, 4].

In Sect. 2.6 a second method for the harmonic oscillator is given bypassing LPBC and LPEQC. It will be seen that the second method is technically much simpler to use as compared to the first method.

We will show how QHJ leads to the eigenvalues without need for computation of the wave function. Knowing the eigenvalues, QHJ can then be used to determine the wave functions.

This second approach bypasses both LPBC and evaluation of contour integral appearing in LPEQC.

However, the original approach Leacock and Padgett, explained in Sect. 2.5, will become unavoidable in treatment of a potential with two phases of supersymmetry in Sect. 4.3.

© The Author(s), under exclusive license to Springer Nature Switzerland AG 2022   9
A. K. Kapoor et al., *Quantum Hamilton-Jacobi Formalism*,
SpringerBriefs in Physics, https://doi.org/10.1007/978-3-031-10624-8_2

2.2 Results from Theory of Complex Variables

Most of the text books will have discussion of the results needed here, see, for example, Churchill [5] and Titchmarsh [6].

2.2.1 Laurent Expansion

A special case of the Laurent expansion theorem is stated below. This result will be used repeatedly in later parts of the book.

Theorem 2.1 *Let $f(z)$ be a function of a complex variable z having an isolated singular point z_0. Let C be a closed contour such that $f(z)$ is analytic inside and on the contour C and z_0 is the only singular point enclosed by the contour. It follows, from the Laurent expansion theorem, that $f(z)$ can be expanded in a series of positive and negative powers of z*

$$f(z) = \sum_{n=1}^{\infty} \frac{b_n}{(z-z_0)^n} + a_0 + \sum_{n=1}^{\infty} a_n(z-z_0)^n, \qquad (2.1)$$

where the coefficients a_n and b_n are given by

$$a_n = \frac{1}{2\pi i} \int_C \frac{f(z)}{(z-z_0)^{(n+1)}} dz, \qquad (2.2)$$

$$b_n = \frac{1}{2\pi i} \int_C f(z)(z-z_0)^{(n-1)} dz. \qquad (2.3)$$

The sum involving the negative powers of $(z - z_0)$ is called the singular part of the function $f(z)$. In all applications in this book, the series with negative n will terminate and we will need only one or two of the coefficients a_n, b_n in the expansion which will be determined by making use of the QHJ equation.

2.2.2 Liuoville Theorem

Theorem 2.2 *(Liuoville) If a function $f(z)$ of a complex variable z is analytic every-where and is bounded at infinity, then $f(z)$ is a constant. A function, which is analytic everywhere and which for large $|z|$ grows like $|z|^n$ for some integer n, is a polynomial of degree less than or equal to n.*

For example, if the only singular point of a function $f(z)$ is a simple pole at z_0, with residue b_1, and if it grows like $|z|^N$ for large $|z|$, we have

$$f(z) = \frac{b_1}{(z - z_0)} + g(z) \qquad (2.4)$$

where $g(z)$ is analytic everywhere and must be a polynomial in z of degree N.

2.2.3 Meromorphic Function

A function of a complex variables z is called meromorphic if all its singular points are poles.

The form of a meromorphic function is completely determined by its singular part and its behaviour at infinity.

A special case, applicable to a meromorphic function with only simple poles, can be stated as follows.

Proposition *If $f(z)$ is meromorphic function having only simple poles at z_k, $k = 1, 2, ..n$ and is bounded by a power $|z|^N$, the form of $f(z)$ can be written down immediately as*

$$f(z) = \sum_{k}^{n} \frac{\rho_k}{(z - z_k)} + Q(z) \qquad (2.5)$$

where ρ_k is residue at the pole z_k and $Q(z)$ is a polynomial of degree N.

The above result is adequate for our applications. In fact it will be used to fix the form of QMF and hence determine the energy eigenfunctions.

Proposition *For a meromorphic function, if the point at infinity is an isolated singular point, the number of poles of the function must be finite.*

The arguments leading to the above result are as follows. The point at infinity being an isolated singular point implies that there exists a circle $|z| = R$ such that all singular points lie inside the circle. Assuming that the number of poles is infinite, use of Bolzano-Weierstrass theorem leads to the fact that the poles will have an accumulation point which will be an essential singularity. This contradicts the assumption that the function is meromorphic and its only singular points are poles.

For all models studied in this book, the QMF turns out to be a rational function after a change of variable. A rational function will have a finite number of poles and the point at infinity will be an isolated singular point. We now state an important result for rational functions.

Proposition *For a rational function the sum of all residues, including the residue at infinity, is zero.*

For all models studied in this book, it will not be necessary to evaluate the contour integral appearing in LPEQC. For most of the applications, the above result will be used, instead of LPEQC, to determine the exact energy eigenvalues.

2.3   Results from Theory of Differential Equations

2.3.1   Solutions of Riccati Equation

We begin this section by reminding the reader that the first step in this formalism is a transition from the Schrodinger equation to QHJ equation.

Since the QHJ equation is of the form of a Riccati equation, we will summarize important results about Riccati equation.

We will state and explain the results, in a form ready to be used in later parts of the book. These results follow from the general properties of the Riccati equation and are well known in the literature. For full details we refer the reader to Ince Chap. 4 [7], Hille Chap. 4 [8], Piaggio Chap. 15 [9] and Chap. 2 in [10]. For an excellent general survey of Riccati equation and its applications, we refer the reader to Reid [11].

It is important to point out that the correspondence between Riccati and second-order linear differential equations is not one to one. There correspond an infinite number of linear ordinary differential equations to a given Riccati equation and vice versa. See, for example, Hille Chap. 4 [8], Piaggio, Chap. 15 [9].

Not all the results collected in this section will used in later parts of the book. They have been collected here with an eye for future investigations a reader may want to take up. One such example, where methods of constructing a second solution of Sect. 2.3.6, is likely to be useful, is investigation of the eigenfunctions corresponding to continuous and doubly degenerate energy eigenvalues.

2.3.2   Singular Points of Solution of Riccati Equation

Fixed Singular Points

Consider the Riccati equation of the form of QHJ equation for a potential problem in one dimension.

$$p^2(x) - i\hbar\frac{dp}{dx} + V(x) - E = 0. \tag{2.6}$$

We assume that $V(x)$ has only isolated singular points in the complex x- plane . The singular points of $V(x)$ will be reflected as singular points of the QMF and are known as the *fixed singular points* of QMF $p(x)$.

Moving Singular Points

Besides these fixed singular points, the solution of a Riccati equation can also have a singularity at a point where $V(x)$ is analytic. Such singular points are called *moving singular points*.

As an example of appearance of moving poles, consider the equation

$$p^2 + k\frac{dp}{dx} + k^2 = 0. \tag{2.7}$$

where k is a constant. Note that even though $V(x) = k^2$ is analytic everywhere, (2.7) has a solution $p(x) = k \cot(x - c)$ and the solution $p(x)$ has moving poles at $x = n\pi + c$. It is seen that the position of the pole depends on constant of integration c.

The location of the singular points in this example depends on the initial condition on the solution, and it moves when the initial condition is changed. Such poles are therefore known as moving poles.

The moving poles will be simple poles with residue one. A discussion of fixed and moving singular points and the absence of moving branch points is given in Ince, Sec. 12.51. It is proved that for the absence of branch points as moving singular points, it is necessary that the differential equation must have the form of Riccati equation.

The moving poles of solution of QHJ equation, for energy eigenvalues, are located at the position of the zeros of the wave function. Therefore, we now summarize the results on location of zeros of wave function in the complex plane.

2.3.3 Real and Complex Zeros of the Wave Function

The QHJ formalism relies extensively on the information about the singularities of the QMF. The poles of QMF in the complex plane correspond to the zeros of the solution of the Schrodinger equation as can be seen from the definition $p(x) = -i\hbar\frac{\psi'(x)}{\psi(x)}$ of QMF. These results are well known in mathematics literature. The primary references are Ince, Chapters X, XI, XII and XXI [7] and Hille, Chap. 8 [8].

Theorem 2.3 *Consider the Schrodinger equation for a potential problem in one dimension.*

$$-\frac{\hbar^2}{2m}\frac{d^2\psi(x)}{dx^2} + V(x)\psi(x) = E\psi(x) \tag{2.8}$$

Assume that the potential supports bound states. Then

1. *The eigenvalues corresponding to the bound states are non-degenerate. This means that the eigenfunction can be taken as a real function of x.*
2. *Let the eigenvalues E_n be arranged in increasing order. The eigenfunction corresponding to the lowest energy has no node; the next eigenfunction has one node. In general, the $(n+1)$th eigenfunction has n nodes.*
3. *For all bound states the wave function vanishes at the nodes. These zeros are located on the real line.*

In addition the wave function may have zeros in the complex plane. Corresponding to every zero, real and complex, of the wave function the QMF will have a simple pole. Thus there are n simple poles of the QMF on the real line.

Complex Zeros of the Wave Function

The zeros of wave function correspond to the poles of QMF. In addition to the information about the real zeros, the results about the zeros of the wave function in the complex plane are also needed. The details about complex zeros and will not be presented here for lack of space. Apart from the books mentioned above, an interested reader will find the appendix, of the original paper by Leacock and Padgett, very helpful.

2.3.4   Riccati Equation—Some General Results

The general form of the Riccati equation is

$$p' = A(x)p^2 + B(x)p + C(x).\tag{2.9}$$

where A, B, C are given functions of independent variable x. A general fractional linear transformation is

$$\tilde{p} = \frac{\alpha(x)p + \beta(x)}{\gamma(x)p + \delta(x)},\tag{2.10}$$

where arbitrary functions $\alpha(x), \beta(x), \gamma(x)$ and $\delta(x)$, The transformed function $\tilde{p}$ satisfies a Riccati equation (2.9) of the form

$$\tilde{p}' = \tilde{A}\tilde{p}^2 + \tilde{B}\tilde{p} + \tilde{C}.\tag{2.11}$$

A general fractional linear transformation in (2.10), can be written as a combination of successive special transformations known as inversion, translation and scaling which are defined below.

(a) Inversion $\tilde{p} = \frac{1}{p}$.
(b) Translation $\tilde{p} = p + \alpha(x)$.
(c) Scaling $\tilde{p} = \beta(x)p$.

The transformation property of Riccati equation under these three is summarized in Table 2.1 .

Table 2.1 Inversion, translation, and scaling

	$\tilde{A}$	$\tilde{B}$	$\tilde{C}$
$\tilde{p} = 1/p$	$-C$	$-B$	$-A$
$\tilde{p} = p + \alpha$	A	$B - 2\alpha A$	$C + A\alpha^2 - B\alpha + \alpha'$
$\tilde{p} = \beta(x)p$	A/β	$B + \beta'/\beta$	C

2.3.5 Reduction to a Second-Order Linear Differential Equation

We know that the QHJ is a special from of the Riccati equation. Here we give the relationship between the general form of the Riccati equation, (2.9) to a second-order linear differential equation.

If we define ϕ by $p = -\frac{\phi'}{A(x)\phi}$, the Riccati equation

$$\frac{\mathrm{d}p}{\mathrm{d}x} = A(x)p^2 + B(x)p + C(x) \tag{2.12}$$

gets transformed into a second-order ordinary differential equation. Thus

$$p'x = -\frac{\phi''}{A\phi} + \frac{\phi'^2}{A\phi^2} + \frac{A'\phi'}{A^2\phi}. \tag{2.13}$$

Using expressions for p and p' in (2.12) we obtain a linear differential equation for ϕ

$$A\frac{d^2\phi}{dx^2} - (A' + AB)\frac{d\phi}{dx} + CA^2\phi = 0. \tag{2.14}$$

Cross ratio

If $p_k(x), k = 1, ...4$ are four particular solutions of a Riccati equation, then their cross ratio is a constant.

$$\frac{(p_1 - p_2)(p_3 - p_4)}{(p_1 - p_3)(p_2 - p_4)} = \text{constant, say } C. \tag{2.15}$$

2.3.6 Most General Solution from Given Solution(s)

If a particular solution of a Riccati equation is known, the most general solution of Riccati equations can be constructed.

Using Three Known Particular Solutions

If we know three particular solutions, p_1, p_2, p_3, of a Riccati equation

$$p'(x) = A(x)p^2(x) + B(x)p(x) + C(x). \tag{2.16}$$

a general solution $p(x)$ can be immediately written down by substituting the three solutions for p_1, p_2, p_3 in (2.15) and solving for $p_4(x)$.

When Two Solutions are Known

Let p_1, p_2 be two particular solutions of Riccati equation (2.16). Then we have

$$p_1'(x) = A(x)p_1^2(x) + B(x)p_1(x) + C(x), \tag{2.17}$$
$$p_2'(x) = A(x)p_2^2(x) + B(x)p_2(x) + C(x). \tag{2.18}$$

Subtracting (2.17) and (2.18) from (2.16) respectively, we get

$$p' - p_1' = A(x)(p^2 - p_1^2) + B(x)(p - p_1), \tag{2.19}$$
$$p' - p_2' = A(x)(p^2 - p_2^2) + B(x)(p - p_2). \tag{2.20}$$

Dividing (2.19) by $(p - p_1)$ and (2.20) by $(p - p_2)$ and subtracting the resulting equations, one is led to

$$\frac{p' - p_1'}{p - p_1} - \frac{p' - p_2'}{p - p_2} = A(x)(p_1 - p_2), \tag{2.21}$$

$$\frac{d}{dx} \log \left(\frac{p - p_1}{p - p_2} \right) = A(x)(p_1 - p_2). \tag{2.22}$$

Integrating we get

$$\log \left(\frac{p - p_1}{p - p_2} \right) = \int A(x)(p_1 - p_2)\, dx + K. \tag{2.23}$$

where K is a constant of integration. Thus we see that one quadrature is required to get the most general solution.

When One Solution is Known

Let $q(x)$ be a particular solution of a given Riccati equation (2.16). Then

$$\frac{dq}{dx} = A(x)q^2 + B(x)q + C(x). \tag{2.24}$$

Writing the required solution p as $p = q + y$ we get the following equation for y

$$\frac{dy}{dx} = A(x)y^2 + \big(2A(x)q + B(x)\big)y, \tag{2.25}$$

$$\text{or} \quad \frac{dy}{dx} = A(x)y^2 + G(x)y, \tag{2.26}$$

where $G(x) = \big(2A(x)q + B(x)\big)$. Dividing the above equation by y^2 and defining $w(x) \equiv 1/y$ we get the following linear equation for $w(x)$:

$$\frac{dw}{dx} + G(x)w(x) + A(x) = 0. \tag{2.27}$$

This equation can be solved for $w(x)$ by making use of $\exp\left(\int G(x)\,dx\right)$ as an integrating factor. Thus we get the answer for $w(x)$ in terms of two quadratures given by

$$w(x) = -e^{-\int dx\, G(x)} \int e^{\int dx\, G(x)} A(x) dx. \tag{2.28}$$

2.4 Some Frequently Used Results and Examples

In this section we derive results on residue at moving poles, change of variable and on QHJ equation in the neighbourhood of infinity. These results will be required in intermediate steps of computation for many problems in later chapters.

2.4.1 Residue of QMF at a Moving Pole

Consider the QHJ equation

$$p^2 - i\hbar\frac{dp}{dx} + 2m\big(V(x) - E\big) = 0. \tag{2.29}$$

Assume that $x = x_0$ is a moving pole of $p(x)$ in complex x- plane. By definition of moving pole, $V(x)$ is analytic at $x = x_0$. We will now show that a pole at such point will be of first order only and will have residue $-i\hbar$.

Assume that a solution $p(x)$ has a pole of order n at $x = x_0$, then its Laurent expansion at $x = x_0$ will have the form

$$p(x) = \frac{b_n}{(x - x_0)^n} + \frac{b_{n-1}}{(x - x_0)^{n-1}} + \cdots + \frac{b_1}{x - x_0} + \cdots + a_0 + a_1(x - x_0) + \cdots .$$

(2.30)

The function $V(x)$ has a Taylor expansion in positive powers of $x - x_0$, and we write it in the form

$$V(x) = c_0 + c_1(x - x_0) + \frac{c_2}{2!}(x - x_0)^2 + \cdots .$$

(2.31)

Substituting the above series expansion in the QHJ equation (2.29), we get

$$2mE = \left\{ \frac{b_n}{(x - x_0)^n} + \frac{b_{n-1}}{(x - x_0)^{n-1}} + \cdots + \frac{b_1}{(x - x_0)} + \cdots \right\}^2$$
$$+ i\hbar \left\{ n\frac{b_n}{(x - x_0)^{n+1}} + (n - 1)\frac{b_{n-1}}{(x - x_0)^n} + \cdots + \frac{b_1}{(x - x_0)^2} + \cdots \right\}$$
$$+ i\hbar(a_1 + 2a_2(x - x_0) + \cdots) + 2m(c_0 + c_1(x - a) + \cdots). \quad (2.32)$$

Starting with $(x - x_0)^{-2n}$, we comparing successive powers of $(x - x_0)$. For purpose of residue at $x = x_0$, it is sufficient to look at negative powers of $(x - x_0)$. If $n > 1$, the highest negative power $2n$ comes only from the series inside the first set of curly brackets and has the coefficient b_n^{2n}. This gives $b_n = 0$, if $n > 1$. The coefficient of $1/(x - x_0)^2$, coming from $n = 1$ terms inside the two curly brackets, when equated to zero gives

$$b_1^2 + i\hbar b_1 = 0 \Rightarrow b_1 = 0, -i\hbar. \quad (2.33)$$

The case $b_1 = 0$ does not correspond to a pole; this proves that a moving pole is always a simple pole with residue $-i\hbar$. Note that this result has been proved only for an equation of form (2.29).

2.4.2 Evaluating Action Integral

In this section we give a result about the evaluation of the action integral in terms of singularities of the QMF outside the contour appearing in the definition of the action integral.

We work in the extended complex plane and define the concepts of an isolated singular point, pole of order m for the point at infinity. Given a function $f(z)$, any statement about the behaviour of the function at infinity will be given a meaning by looking at the behaviour of the function $f(1/\zeta)$ at $\zeta = 0$. So for example, we say $f(z)$ has a pole of order m at infinity if $f(1/\zeta)$ has a pole of order m at $\zeta = 0$. For a discussion of analytic properties of a function at infinity and other related issues, see Sect. 6.6 of [12].

In this section we will define the residue of a function $f(z)$ at infinity and give a prescription, see a boxed text below, for evaluating this residue. This result will be needed in the next section and in later chapters.

Let a function $f(z)$ be such that it has an isolated singular point at infinity. The residue of the function $f(z)$ at infinity is defined as, see, for example, Sect. 6.6 [12],

$$\text{Res}\{f(z)\}|_\infty \overset{\text{def}}{=} -\frac{1}{2\pi i} \oint_C f(z)dz. \tag{2.34}$$

where C is an anticlockwise closed contour such that the function $f(z)$ is analytic everywhere on and outside the contour C, except possibly at infinity, see Fig. 2.1. In order to be able to compute the residue at infinity, a scheme of evaluating the integral, in the right hand side of (2.34), is required.

We draw a positively oriented circle $|z| = R$ of sufficiently large radius R and enclosing the contour C completely, see Fig. 2.1a. Then we have

$$\oint_C f(z)dz = \oint_{C_R} f(z)dz. \tag{2.35}$$

Thus out side C_R the function $f(z)$ is analytic in complex z- plane, except at infinity. Next we use the mapping $z \to \zeta = 1/z$ and transform the integral in the z- plane into an integral in complex ζ- plane. The anticlockwise circle C_R in the complex plane is mapped into a small *clockwise circle*, γ, of radius $1/R$ in the complex ζ- plane as in Fig. 2.1b.

Thus we get

$$\oint_C f(z)dz = \oint_{C_R} f(z)dz = -\oint_\gamma (\tilde{f}(\zeta)/\zeta^2)d\zeta \tag{2.36}$$

where $\tilde{f}(\zeta) = f(z)|_{z \to 1/\zeta}$. The last integral in the above is now evaluated by applying the residue theorem.

The only singular point inside the circle γ in the complex ζ- plane is $\zeta = 0$, and therefore we get

Fig. 2.1 Mapping for evaluation of integral $\oint_{C_R} f(z)\,dz$

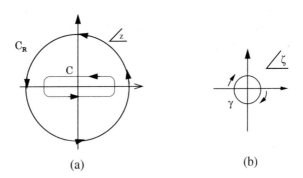

(a)   (b)

$$\oint_C f(z)dz = 2\pi i \times \text{Res}\{\tilde{f}(\zeta)/\zeta^2\}\Big|_{\zeta=0}. \tag{2.37}$$

Note that here an additional negative sign appears because the contour γ is clockwise. The residue in the above equation is just the coefficient of ζ in expansion of $\tilde{f}(\zeta)$ in powers of ζ. Comparing Eq.(2.37) with the definition (2.34), we arrive at the following prescription

> The residue of $f(z)$ at infinity is given by negative of the coefficient of ζ in the Laurent expansion of $f(1/\zeta)$ in powers of ζ.

If the function $f(z)$ has isolated singular points outside the contour C, the value of the $\oint_C f(z)\,dz$ will be given by negative of sum of residues at all the singular points outside C, including the point at infinity.

The above result is useful to evaluate the contour integral in appearing in the action integral by adding up contributions of the singular points of the QMF outside the contour in LPEQC. For more details, and for examples on residue at infinity, see Sect. 6.6 [12].

2.5 Harmonic Oscillator—Method-I

In this section we solve the harmonic oscillator problem following the QHJ scheme making use of the LPBC and LPEQC. The treatment of harmonic oscillator given in this section was reported in [1, 2] where solutions to several other models can also be found.

In the next section we will give an alternate and simpler solution for the harmonic oscillator problem without using LPBC and LPEQC.

2.5.1 Behaviour of p_{cl} for Large $|x|$

In our approach the form of QMF is determined by an application of Liuoville's theorem. One of the inputs for Liuoville's theorem is the behaviour of QMF for large $|x|$. We will now show that, for all cases of our interest, this behaviour can be guessed by dropping dp/dx term in QHJ equation

$$p^2 - i\hbar\frac{dp}{dx} + 2m(V(x) - E) = 0. \tag{2.38}$$

The method of solution of QHJ works because QMF turns out to be a rational function, i.e., a function of the form $P(y)/Q(y)$, where $P(y)$, $Q(y)$ are polynomials in some variable y. The QHJ equation then implies that the potential like term in the QHJ equation in the new variable must also be a rational function. A discussion of properties of a function for large $|x|$, in the extended complex x- plane, is best done by looking at its behaviour near $\zeta = 0$ where $\zeta = 1/x$.

We start with the assumption that the QMF $p(x)$ has a pole of some order M at infinity. This assumption is found to be true for all cases studied in this monograph. This means that $\tilde{p}(\zeta)$ has a pole of order M at $\zeta = 0$. Therefore, we can write the Laurent expansion of $f(\zeta)$ about $\zeta = 0$

$$\tilde{p}(\zeta) = \sum_{n=1}^{M} b_n \zeta^{-n} + \sum_{n=0}^{\infty} a_n \zeta^n \qquad (2.39)$$

converging in a neighbourhood of $\zeta = 0$. In terms of the variable x, we have an expansion

$$p(x) = \sum_{n=1}^{M} b_n x^n + \sum_{n=0}^{\infty} a_n x^{-n}. \qquad (2.40)$$

converging outside a circle $|x| > R$, for some R. Also we have

$$\frac{dp}{dx} = \sum_{n=1}^{M} n b_n x^{n-1} + \sum_{n=0}^{\infty} (-n) a_n x^{-n-1}. \qquad (2.41)$$

If the potential $V(x)$ is a ratio of two polynomials, it will also have a Laurent expansion similar to $p(x)$ as in (2.45).

The behaviour of the QMF for large $|x|$ can now be determined from the QHJ equation

$$p^2 - i\hbar \frac{dp}{dx} + 2m \left(\frac{1}{2} m\omega^2 x^2 - E \right) = 0. \qquad (2.42)$$

Substituting the expansions (2.40) and (2.41) in Eq.(2.42), we demand that the coefficients of each power of x be zero. For our present purpose, we only need to check the positive powers of x. The highest positive powers of x in the three terms are x^{2M}, x^{M-1} and x^2, respectively. For $M \geq 2$, the highest power of x is greater than or equal to four and comes from the term p^2 only. Therefore, the absence such terms requires that $M \leq 1$. For $M = 1$ the highest power is x^2 and equating the coefficient of x^2 to zero gives $b_1 = \pm im\omega x$. Thus we have

$$b_n = \begin{cases} 0, & \text{if } n > 1 \\ \pm im\omega x & \text{for } n = 1. \end{cases} \qquad (2.43)$$

and for large $|x|$, we have

$$p(x) = \pm im\omega x + \cdots \qquad (2.44)$$

The arguments presented above can also be extended to the cases where the potential term is a rational function of x. This will involve an extra step of expanding the potential also in a Laurent expansion. A closer look at the above analysis justifies dropping of dp/dx to obtain the leading term for large $|x|$ in the QMF. Note that this justification applies only when the point at infinity is an isolated singular point.

2.5.2   Classical Momentum in the Complex Plane

Taking oscillator potential to be $V(x) = \frac{1}{2}m\omega^2x^2$, the classical momentum is given by

$$p_{cl} = \sqrt{2m\left(E - \frac{1}{2}m\omega^2x^2\right)}. \tag{2.45}$$

There are two turning points at $x = \pm a$, where $a = \sqrt{2E/m\omega^2}$. To simplify the notation we take the classical momentum as

$$p_{cl} = K\sqrt{a^2 - x^2} = iK\sqrt{x^2 - a^2}, \quad \text{where} \quad K = m\omega. \tag{2.46}$$

As a function of complex variable x, the classical momentum is a multivalued function with branch points at $x = \pm a$ and a careful definition is required. For a detailed account of multivalued functions we refer the reader to Ch3. of [12]. For reasons of emphasis, we replace x by z commonly used for complex variables. In the rest of this section x will denote the real part of z.

We define variables $\rho_1, \rho_2, \phi_1, \phi_2$ by

$$z - a = K\rho_1 e^{i\phi_1}, \quad 0 < \phi_1 < 2\pi \tag{2.47}$$
$$z + a = K\rho_2 e^{i\phi_2}, \quad 0 < \phi_2 < 2\pi. \tag{2.48}$$

and $\rho_1 + \rho_2 > 2a$.

We now define the classical momentum function p_{cl} by means of equation

$$p_{cl}(z) = i\lambda K\rho_1^{1/2}\rho_2^{1/2}e^{i\phi_1/2}e^{i\phi_2/2}. \tag{2.49}$$

where $\lambda = \pm 1$ is a constant to be chosen in accordance with Leacock-Padgett specification of the branch for the classical momentum. The ranges for the angles ϕ_1, ϕ_2 correspond to a branch cut of p_{cl} running from $-a$ to a and the classical momentum function is discontinuous for $-a < x < a$.

2.5.3   Boundary Condition on QMF

The classical momentum function is double valued and the Leacock and Padgett specification requires that p_{cl} be defined so that for a point just below the branch cut,

Fig. 2.2 Definition of
$\rho_1, \rho_2, \phi_1, \phi_2$

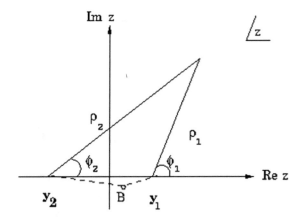

i.e., for $z = x - i\epsilon$, $-a < x < a$, the value of p_{cl} should be positive. From Fig. 2.2, for a point B just below the cut, we have $\phi_1 = \pi$, $\phi_2 = 2\pi$ and therefore Using the definition (2.49), we get a value

$$p_{cl}\big|_B = K\lambda\rho_1^{1/2}\rho_2^{1/2}. \tag{2.50}$$

at a point B, just below the cut. Thus we must choose $\lambda = 1$. For large, positive and real x we have $\phi_1 = 0$, $\phi_2 = 0$, therefore the definition, Eq.(2.49), with $\lambda = 1$ implies that the choice $p_{cl} = +im\omega x$, for large $|x|$, is consistent with LPBC.

Having obtained the large $|x|$ behaviour of QMF, and using the fact that there are no fixed singular points, we can write down the form of QMF as a function of complex variable x:

$$p(x) = \sum_{1}^{n} \frac{-i\hbar}{x - x_k} + im\omega x + C. \tag{2.51}$$

where C is a constant.

2.5.4 Energy Eigenvalues

The action integral appearing in LPEQC

$$\frac{1}{2\pi} \oint p(x)dx = n\hbar \tag{2.52}$$

is over a thin rectangular contour C shown in Fig. 1.1. The contour C encloses all the moving poles corresponding to the nodes of the wave function. For harmonic oscillator, the QMF has no singularity in the complex plane outside the contour C. Hence as discussed in Chap. 1 Sect. 2.4.2 , the contour integral is given by

$$\oint_C p(x)dx = -(2\pi i)\text{Res}\{p(x)\}\Big|_\infty.$$

To obtain the residue at infinity, we first change the variable from x to $\zeta = 1/x$ in QHJ equation.

Changing the variable from x to $\zeta = 1/x$, the quantum Hamilton-Jacobi equation,

$$p^2(x) - i\hbar \frac{dp(x)}{dx} + ((m\omega x)^2 - 2mE) = 0 \tag{2.53}$$

takes the form

$$\tilde{p}(\zeta)^2 + i\hbar\zeta^2 \frac{d}{d\zeta}\tilde{p} + \frac{m^2\omega^2}{\zeta^2} - 2mE = 0 \tag{2.54}$$

We expand $\tilde{p}$ in the Laurent expansion in the neighbourhood of $\zeta = 0$

$$\tilde{p}(\zeta) = \frac{im\omega}{\zeta} + a_0 + a_1\zeta + \dots. \tag{2.55}$$

Here the first term has been written down after taking into account of the leading behaviour $im\omega x$, obtained earlier, see Eq.(2.51). Next we substitute the expansion of $\tilde{p}(\zeta)$ in (2.54) and collect different powers of ζ. The ζ^2 terms cancel, and coefficient of ζ^{-1} when equated to zero gives $a_0 = 0$. Continuing in this way, a_1 is determined by equating the term independent of ζ to 0. This gives

$$2im\omega a_1 + \hbar m\omega - 2mE = 0 \Longrightarrow a_1 = -i\left(\frac{E}{\omega} - \frac{\hbar}{2}\right). \tag{2.56}$$

The residue of $p(x)$ at infinity is given by

$$\text{Res}\{p(x)\}\Big|_\infty = -\text{coeff of } \zeta \text{ in the expansion of } \tilde{p}(\zeta) \tag{2.57}$$

$$= -a_1. = i\left(\frac{E}{\omega} - \frac{\hbar}{2}\right) \tag{2.58}$$

Thus the integral in the left hand side of (2.52) becomes

$$\oint_C p(x)\,dx = (-2\pi i)\text{Res}\{p(x)\}\Big|_\infty = (2\pi)\left(\frac{E}{\omega} - \frac{\hbar}{2}\right). \tag{2.59}$$

Recall that a negative sign appears as the point at infinity is outside the positively oriented contour C. Thus the quantization condition (2.52) gives the energy eigenvalues as

$$E = (n + 1/2)\hbar\omega. \tag{2.60}$$

Here we have demonstrated how the complex contour integral appearing in the action integral can be computed to give the energy eigenvalues. This method was followed in references [1, 2] to obtain the energy eigenvalues for several ES models.

In the next section we will present an alternative, simpler method of finding the eigenvalues and eigenfunctions for the harmonic oscillator problem.

2.6 Harmonic Oscillator—Method-II

Here we use square integrability requirement to choose the correct behaviour of QMF at infinity, see Eq. (2.44), in place of LPBC. Also we show how QHJ equation can be used to determine the energy eigenvalues in a very simple manner. This alternative approach is much simpler as compared to evaluation of the contour integral $J(E)$ appearing in the LPEQC. Once the energy eigenvalues have been determined the eigenfunctions can be found in a straight forward manner making use of the QHJ equation.

2.6.1 Using Square Integrability

In Sect. 2.5.1 it was proved that the behaviour of QMF for large $|x|$, see Eq.(2.44), is given by

$$p(x, E) \approx \pm im\omega x + 0(x^0). \tag{2.61}$$

Among these, the physically acceptable solution was identified by applying the LPBC. However, in this case, it is much simpler to choose the correct sign from (2.61) by demanding that the wave function for large and real x goes to zero, as $x \to \pm\infty$.

The behaviour of the wave function $\psi(x)$ as $x \to \pm\infty$ is given by

$$\psi(x) = \exp\left(\frac{i}{\hbar} \int p dx\right) \approx \exp\left(\mp \frac{m\omega x^2}{2\hbar}\right). \tag{2.62}$$

In order that the wave function, for real x, may not blow up as $x \to \pm\infty$ infinity, we choose

$$p(x, E) \approx im\omega x + 0(x^0). \tag{2.63}$$

The above equation gives the desired behaviour of QMF for large x.

2.6.2   General Form of QMF

The bound state wave function corresponding to the nth excited state has n zeros on the real axis. The QMF will have n simple poles, known as the moving poles, with residue $-i\hbar$, (see Sect. 2.4.1), at the points corresponding to the zeros of the wave function. Since the potential function $V(x) = \frac{1}{2}m\omega^2 x^2$ is an entire function, it turns out that $p(x, E)$ does not have any other singular points. Thus with no fixed singular point, and n moving poles, the QMF is a meromorphic function and has a representation, (see Proposition 1 in Sect. 2.2.2),

$$p(x, E) = \sum_{k=1}^{n} \frac{-i\hbar}{(x - x_k)} + im\omega x + c, \tag{2.64}$$

where c is a constant.

2.6.3   Energy Eigenvalues

We substitute the expression (2.64) in QHJ equation

$$p^2 - i\hbar \frac{d}{dx} + 2m\left(\frac{1}{2}m\omega^2 x^2 - E\right) = 0. \tag{2.65}$$

and we get

$$\left(\sum_{k=1}^{n}\left(\frac{-i\hbar}{x - x_k}\right) + im\omega x + c\right)^2 - i\hbar\left(\sum_{k=1}^{n} + \frac{i\hbar}{(x - x_k)^2} + im\omega\right)$$
$$+ m^2\omega^2 x^2 - 2mE = 0 \tag{2.66}$$

Expanding all brackets, we see that $\frac{1}{(x-x_k)^2}$ terms cancel. Next we expand $\frac{1}{x-x_k}$ in a Laurent series in a neighbourhood at infinity, i.e. in the region where $|x| > |x_k|$, for all k. This expansion is given by

$$\frac{1}{x - x_k} = \frac{1}{x}\frac{1}{[1 - (x_k/x)]} = \frac{1}{x}\left(1 + \frac{1}{x_k} + \frac{1}{x_k^2} + \cdots\right) \tag{2.67}$$

and converges for $\{x | x > x_k, \text{all } k\}$. Substituting (2.67) in (2.66), the leading terms in the expansion of (2.66) for large $|x|$ are retained and coefficients of different powers of x in the l.h.s. are equated to zero. After some algebra it can be seen that

- the x^2 terms cancel,
- terms linear in x have coefficient $\frac{2m\omega}{\hbar}(ic)$ and hence c must be zero,
- terms independent of x, when equated to zero, give

$$2im\omega(-in\hbar) - i\hbar im\omega - 2mE = 0.$$

This leads to the desired energy eigenvalues

$$E = \left(n + \frac{1}{2}\right)\hbar\omega. \tag{2.68}$$

2.6.4 Energy Eigenfunctions

The energy eigenfunctions are now easily obtained as follows. We note that

$$\sum_{k=1}^{n} \frac{1}{(x - x_k)} = \frac{d}{dx}\ln P_n(x) = \frac{P_n'}{P_n} \tag{2.69}$$

where $P_n(x)$ denotes the nth degree polynomial $\prod_{k=1}^{n}(x - x_k)$. Thus Eq. (2.51) for QMF takes the form

$$p(x, E) = -i\hbar\frac{P_n'(x)}{P_n(x)} + im\omega x. \tag{2.70}$$

Substituting $p(x, E)$ in Eq.(2.65), one gets the following second-order linear differential equation for polynomial $P_n(x)$

$$\frac{1}{\alpha^2}\frac{d^2 P_n(x)}{dx^2} - 2x\frac{dP_n(x)}{dx} + nP_n(x) = 0, \quad \text{where} \quad \alpha = \sqrt{\frac{m\omega}{\hbar}}. \tag{2.71}$$

This is the Hermite equation, a linear second-order differential equation and has two linearly dependent solutions. Its polynomial solution is Hermite polynomial, $H_n(\alpha x)$. Thus we get the final energy eigenfunctions, as

$$\psi_n(x) = C_n H_n(\alpha x)e^{-\alpha^2 x^2/2},$$

where C_n is a constant to be fixed by normalization requirement. From the standard series solution method given in text books it is known that the second solution of (2.71) is an infinite series diverging as $\exp(\alpha^2 x^2)$ and is an unphysical solution,

The second method present here is much more transparent, simple and elegant as compared to the standard series solution method given in most of the text books.

We conclude this chapter by mentioning that only application in Chap. 4 Sect.4.3, it will be necessary to follow the original approach of Leacock and Padgett. For all other applications, an alternative simpler approach, suited to the problem at hand, will be used.

References

1. Bhalla R.S.: The Quantum Hamilton-Jacobi Formalism Approach to Energy Spectra of Potential Problems, Ph. D. Thesis submitted to University of Hyderabad, 1996
2. Bhalla, R.S., Kapoor, A.K., Panigrahi, P.K.: Quantum Hamilton-Jacobi formalism and bound state spectra. Am. J. Phys. **65**, 1187 (1997)
3. Leacock, R.A., Padgett, J.: Hamilton-Jacobi Theory and the quantum action variable. Phys. Rev. Lett. **50**, 3 (1983)
4. Leacock, R.A., Padgett, M.J.: Hamilton-Jacobi Action angle quantum mechanics. Phys. Rev. D **28**, 2491 (1983)
5. Churchill Ruel Vance and Brown James: Complex Variables and Applications. McGraw-Hill Publishing Book Co., New York (2014)
6. Titchmarsh, E.C.: The Theory of Functions. Oxford University Press, United Kingdom (1964)
7. Ince, E.L.: Ordinary Differential Equations. Dover Publications, New York (1956)
8. Einar, Hille: Ordinary Differential Equations in Complex Domain. Wiley Interscience Publications, New York (1976)
9. Piaggio, H.T.H.: An Elementary Treatise on Differential Equations and Their Applications. B. Bell and Sons, London (1949)
10. Shabat, A., Kartashova, E.: Computable Integrability. arXiv:1103.2423v1
11. Reid, W.T.: Riccati Differential Equations. Academic Press, New York and London (1972)
12. Kapoor, A.K.: Complex Variables—Principles and Problem Sessions, World Scientific Publishing Co. Pte. Ltd., Singapore (2011); For errata, visit. https://0space.org/node/3488

Chapter 3
Exactly Solvable Models

3.1 Introduction

In Chaps. 1 and 2 the Leacock-Padgett QHJ [1, 2] scheme has been presented along with the mathematical preparation required for application to potential models. Taking the example of the harmonic oscillator, we also illustrated how the eigenvalues and eigenfunctions can be obtained using the QHJ scheme.

In this chapter we present a complete solution for the bound state energy eigenvalues and eigenfunctions of the Morse oscillator, of the radial oscillator and of the particle in a box. The methods used here, for the Morse oscillator and the radial oscillator, will serve as examples for other potential problems in later chapters. Also the results obtained for the radial oscillator will be needed in chapter five, where a rational extension of this model will be discussed.

The treatment of particle in a box within the QHJ scheme [3, 4] required "some out of the box" thinking due to the fact that the restriction on the range of x is equal to the box size, and also due to the fact that the QHJ scheme is formulated in the entire complex plane. Problems other than the particle in a box make use of a change of real variable from x- to a new real variable. The problem of particle in a box is a unique problem in the sense that it required use of a conformal mapping from x-plane to a complex variable $z = e^{2\pi i x/L}$. The energy levels have been obtained by transforming the LPEQC contour integral from the complex x- plane to another contour in the complex z- plane. In Sect. 3.6 a set of conditions required for exactness of supersymmetric WKB approximation (SWKB) is given in the context of QHJ.

3.2 Change of Variable

A first step in obtaining solution to many quantum mechanical problems, including the Morse oscillator, is a change of the independent variable. We describe a sequence

© The Author(s), under exclusive license to Springer Nature Switzerland AG 2022
A. K. Kapoor et al., *Quantum Hamilton-Jacobi Formalism*,
SpringerBriefs in Physics, https://doi.org/10.1007/978-3-031-10624-8_3

of steps which result in an equation of the same final form in the new variable as the QHJ equation in the old variable.

In this chapter we will work with units so that we have $\hbar = 1$, $2m = 1$. The bound state energy eigenfunctions in one dimension are known to be nondegenerate [5]. Therefore, for real potentials, if $\psi(x)$ is a bound state eigenfunction with energy E, so is $\psi^*(x)$ and these two cannot be linearly independent. Thus $\psi(x) = c\psi^*(x)$ for some complex constant c. This implies that the bound state eigenfunctions can be chosen to be real. Therefore, for application to bound states, we mostly use q defined by

$$q(x, E) = \frac{\psi'(x)}{\psi(x)}, \tag{3.1}$$

which will be real for real x. Thus

$$\psi = \exp\left(\int q(x, E)dx\right); \quad q(x, E) = \frac{d}{dx}\log\psi = \frac{1}{\psi}\frac{d\psi}{dx},$$

so that $q(x, E) = ip(x, E)$ and the equation for $q(x)$ takes the form

$$q^2(x, E) + \frac{dq(x, E)}{dx} + E - V(x) = 0. \tag{3.2}$$

We shall call $q(x, E)$ also as the QMF and (3.2) as the QHJ equation. No confusion should arise as for a given model either $p(x, E)$, or $q(x, E)$ will be used. Also it should be obvious, see Sect.2.4.1, that the residue of q at a moving pole will be unity.

Under a change of variable $y \equiv y(x)$, we have

$$\frac{d}{dx} = \frac{dy}{dx}\frac{d}{dy} \equiv F(y)\frac{d}{dy} \tag{3.3}$$

Here $F(y)$ is the derivative dy/dx expressed as a function of y. The QHJ equation (3.2) takes the form

$$\tilde{q}^2 + F(y)\frac{d\tilde{q}}{dy} + E - V = 0, \tag{3.4}$$

where $\tilde{q}(y, E) = q(x(y), E)$. We perform a sequence of transformations to bring the above equation to the same form as the QHJ equation (3.2). To achieve this, we define

$$\tilde{q}(y, E) = F(y)\chi(y, E) \tag{3.5}$$

so that

$$\tilde{q}' = F'(y)\chi + F(y)\chi' \tag{3.6}$$

where prime denotes derivatives w.r.t. the argument y. QHJ, Eq. (3.4), takes the form

$$F^2 \chi^2 + F^2 \chi' + F F' \chi + \frac{E - V}{F^2} = 0.$$ (3.7)

Divide by F^2

$$\chi^2 + \chi' + \frac{F'}{F} \chi + \frac{(E - V)}{F^2} = 0.$$ (3.8)

and completing the squares, we rewrite the above equation as

$$\left(\chi + \frac{1}{2} \frac{F'}{F}\right)^2 - \frac{1}{4} \left(\frac{F'}{F}\right)^2 + \chi' + \frac{E - V}{F^2} = 0.$$ (3.9)

Defining

$$\phi = \chi + \frac{1}{2} \frac{F'}{F}$$ (3.10)

so that

$$\chi' = \phi' - \frac{1}{2} \frac{F''}{F} + \frac{1}{2} \left(\frac{F'}{F}\right)^2.$$ (3.11)

The QHJ in terms of ϕ takes a simple form

$$\phi^2 + \phi' + \frac{E - V}{F^2} + \frac{1}{4} \left(\frac{F'}{F}\right)^2 - \frac{1}{2} \frac{F''}{F} = 0.$$ (3.12)

In terms of ΔV defined by

$$\Delta \tilde{V} = \frac{1}{2} \left(\frac{F''}{F}\right) - \frac{1}{4} \left(\frac{F'}{F}\right)^2$$

$$= \frac{1}{2} (\log F)'' + \frac{1}{4} [(\log F)']^2.$$

With $\tilde{V} = \dfrac{V - E}{F^2} + \Delta \tilde{V}$, the final form of equation for ϕ becomes

$$\boxed{\phi^2 + \phi' - \tilde{V} = 0}$$ (3.13)

This equation for ϕ is of the same form as that of QHJ equation for q. This form has the advantage that the moving poles of ϕ have unit residue.

Expression of the wave function in terms of ϕ:
For the computations of the eigenfunctions in the later sections, we need the following relation

$$\psi(x) = \exp \left(\int q(x) dx\right).$$ (3.14)

Expressing $\chi(x)$ in terms of ϕ from (3.10), it is easily seen that $\psi(x)$ in terms of ϕ is given by

$$\psi(x) = \frac{1}{\sqrt{F}} \exp\left(\int \phi(y)dy\right). \tag{3.15}$$

3.3   Morse Oscillator

The potential energy of the Morse oscillator, in notation of Cooper et al. [7], is

$$V(x) = A^2 + B^2 e^{-2\alpha x} - 2B\left(A + \frac{\alpha}{2}\right)e^{-\alpha x}, \qquad A, B > 0. \tag{3.16}$$

The QHJ equation is given by

$$p^2(x, E) - ip'(x, E) - \left[E - A^2 - B^2 e^{-2\alpha x} + 2B\left(A + \frac{\alpha}{2}\right)e^{-\alpha x}\right] = 0. \tag{3.17}$$

It is known in literature on supersymmetric quantum mechanics that the potential $V(x)$ can be written as $V(x) = W^2 - W'$ where

$$W(x) = A - B e^{-\alpha x} \tag{3.18}$$

is known as super potential $W(x)$. In terms of the super potential $W(x)$ the Schrödinger equation factorizes as

$$\left(-\frac{d}{dx} + W(x)\right)\left(\frac{d}{dx} + W(x)\right)\psi = E\psi. \tag{3.19}$$

It is obvious from this equation that $\psi_0(x) \equiv \exp\left(-\int W(x)\,dx\right)$ is the energy eigenfunction with corresponding eigenvalue is $E = 0$. This is also the ground state solution. The last statement follows from the fact that potential obeys $V(x) \geq 0$ for all x implying that $E = 0$ is the lowest possible energy. It can now be easily seen that the QMF for the ground state coincides with $W(x)$ apart from an overall constant i:

$$p(x, E = 0) = \frac{-i}{\psi_0(x)}\frac{d\psi_0(x)}{dx} = iW(x) \tag{3.20}$$

We will use this relation in place of LPBC to select physically appropriate value of a residue b_1, see (3.29) below.

Change of variable
Using a transformation to a new independent variable

$$y = \frac{2B}{\alpha} e^{-\alpha x} \tag{3.21}$$

the QHJ equation in the new variable y becomes

$$\tilde{p}^2(y, E) + i\alpha y \tilde{p}'(y, E) - \left[E - A^2 - \frac{\alpha^2}{4} y^2 + \left(A + \frac{\alpha}{2} \right) \alpha y \right] = 0. \quad (3.22)$$

The effect of the change of variable is to convert the original QHJ equation into a Riccati equation with rational coefficients.

Next we perform a sequence of operations on the dependent variable as explained below. This will lead to an equation like QHJ with a rational function replacing the potential term.

First we define $\phi(y, E)$ given by

$$\tilde{p}(y, E) \equiv i\alpha y \phi(y, E), \qquad \phi(y, E) = (-i/\alpha y) \tilde{p}(y, E). \quad (3.23)$$

(3.22) takes the form

$$\left(\phi + \frac{1}{2y} \right)^2 + \phi' - \frac{1}{4y^2} + \frac{1}{\alpha^2 y^2} \left[E - A^2 - \frac{y^2 \alpha^2}{4} + \left(A + \frac{\alpha}{2} \right) y\alpha \right] = 0. \quad (3.24)$$

Next we introduce

$$\chi(y, E) = \phi(y, E) + \frac{1}{2y}, \quad (3.25)$$

and get an equation for χ

$$\chi^2 + \chi' + \frac{1}{4y^2} + \frac{1}{\alpha^2 y^2} \left[E - A^2 - \frac{y^2 \alpha^2}{4} + \left(A + \frac{\alpha}{2} \right) y\alpha \right] = 0, \quad (3.26)$$

having the same form as the original QHJ equation, (3.16), except that a rational function of y appears in place of the potential.

For later use, we note that QMF $\tilde{p}(y, E)$, (3.23), in terms of χ is given by

$$\tilde{p} = i\alpha y \phi(y) = i\alpha y(\chi - 1/(2y)) \quad (3.27)$$

Form of QMF $\chi(y)$

Equation (3.26) suggests that χ has a fixed pole at $y = 0$. It will also have n moving poles corresponding to the nodes of the wave-function. We assume that there are no other poles in the finite complex plane. Moreover, from Eq. (3.26), χ is seen to be bounded for large $|y|$. Hence using Liouville's theorem we get

$$\chi = \frac{b_1}{y} + \sum_{k=1}^{n} \left(\frac{1}{y - y_k} \right) + c, \quad (3.28)$$

where b_1, c and y_k are constants to be fixed. Here the summation term represents the contribution of moving poles and can be written in a compact form as

$$\sum_{k=1}^{n} \frac{1}{y - y_k} = \frac{P'(y)}{P(y)}.$$

Here $P(y)$ is a polynomial of degree n given by

$$P(y) = \prod_{k=1}^{n} (y - y_k).$$

and the function χ takes the form

$$\chi = \frac{b_1}{y} + \frac{1}{P} \frac{dP}{dy} + c. \tag{3.29}$$

To determine the function χ, we need to compute the residue of χ at $y = 0$, the constant c and the polynomial $P(y)$.

Computation of residue at $y = 0$

The residue b_1 at $y = 0$ is determined by substituting the form (3.29) of $\chi(y)$ in Eq. (3.26) for χ and comparing the coefficients of different powers of y. This gives two values of b_1

$$b_1 = \frac{A}{\alpha} + \frac{1}{2}, \qquad b_1 = 1 - \left(\frac{A}{\alpha} + \frac{1}{2}\right). \tag{3.30}$$

Instead of using LPBC to make correct choice of residue, we use the relation (3.20) between the QMF and the superpotential, see (3.20). Since the ground state has no nodes, $\tilde{p}(x, E)$ should coincide with $iW(x)$ for $n = 0$. This leads us to the desired answer

$$b_1 = \frac{A}{\alpha} + \frac{1}{2}. \tag{3.31}$$

Using the value of b_1 from (3.31) and substituting (3.29) in (3.26) gives

$$\frac{P''}{P} + 2\frac{b_1}{y}\frac{P'}{P} + 2\frac{P'}{P}c + 2\frac{b_1}{y}c + c^2 - \frac{1}{4} + \frac{1}{\alpha y}\left(A + \frac{\alpha}{2}\right) = 0 \tag{3.32}$$

We will use this equation in the following section to

(a) fix the constant c,
(b) determine the energy eigenvalues,
(c) derive a differential equation for the polynomial $P(y)$,
(d) obtain full expression for the QMF and the energy eigenfunctions.

Energy eigenvalues

In order to proceed further we look at the behavior of each term for large y. Using the leading terms

$$\frac{P''(y)}{P(y)} \sim \frac{n(n-1)}{y^2}, \qquad \frac{P'(y)}{P(y)} \sim \frac{n}{y}$$

in Eq. (3.32), and equating the constant term on both sides gives, $c = \pm\frac{1}{2}$. The correct sign for c is chosen by the condition of square integrability on the wave function which fixes $c = -\frac{1}{2}$.

Comparing the coefficient of $\frac{1}{y}$ for large y on both sides we get

$$2b_1 c + 2nc + \left(A + \frac{\alpha}{2}\right)\frac{1}{\alpha} = 0, \tag{3.33}$$

which on using the values of b_1 and c and on simplification gives the energy eigenvalue

$$E = A^2 - (A - n\alpha)^2. \tag{3.34}$$

Energy eigenfunctions

Substituting the values of b_1 and c in Eq. (3.32), we get

$$y P''(y) + \{1 - y + 2(s - n)\}\, P'(y) + n P(y) = 0. \tag{3.35}$$

where $s = A/\alpha$. Comparing this with the standard Laguerre differential equation,

$$xy'' + (\beta + 1 - x)y' + ny = 0,$$

we have $P(y) \equiv \text{const } L_n^\beta(y)$.[1]

The wave function for the Morse oscillator is obtained from

$$\psi(x) = \exp\left(i \int p(x, E)\mathrm{d}x\right). \tag{3.36}$$

In terms of the variable y, we have

$$\psi(y) = \exp\left(i \int \left[\frac{b_1}{y} + \frac{P''(x)}{P(x)} - \frac{1}{2} - \frac{1}{2y}\right]\mathrm{d}y\right). \tag{3.37}$$

---

[1] This identification implicitly uses the result that a second-order linear differential equation can have at most one polynomial solution. It cannot have both linearly independent solutions as polynomials. [6].

On integrating and simplifying we get

$$\psi_n(y) = N y^{s-n} \exp(-(y/2)) P(x) \tag{3.38}$$

Here N is a constant of integration to be fixed by normalization. The polynomial $P(y)$ has been already determined to be Laguerre polynomial, $L_\beta^n(y)$, giving

$$\psi_n(y) = N y^{s-n} \exp(-y/2) L_n^\beta(y) \tag{3.39}$$

The expressions for the eigenfunctions agree with the known results in the literature [7]

3.4   Radial Oscillator

In this section, we solve for the eigenvalues and eigenfunctions of the radial oscillator potential using the QHJ formalism. This example will be used to show the application of a different boundary condition to pick the right value of the residues of the QMF. In Chap. 5 the rational extension of radial oscillator will be discussed. All the details that are required in the construction of rational potentials in Chap. 5 are given here. We will follow the notation $\hbar = 1, 2m = 1$.

The radial equation for the harmonic oscillator
We begin with the isotropic oscillator in three dimensions. The corresponding potential is given by

$$V(x, y, z) = \frac{1}{4}\omega^2(x^2 + y^2 + z^2), \tag{3.40}$$

which is an ES and separable problem in three dimensions. The Schrödinger equation is given by

$$-\nabla^2 \Psi(x, y, z) + V(x, y, z)\Psi(x, y, z) = E\Psi(x, y, z), \tag{3.41}$$

with the eigenvalues and eigenfunctions given by [8]

$$E = (2n + \ell + \frac{3}{2})\omega \tag{3.42}$$

and

$$\Psi(r, \theta, \phi) = r^\ell \exp\left(-\frac{1}{4}\omega r^2\right) L_q^{\ell+\frac{1}{2}}\left(\frac{1}{2}\omega r^2\right) P_\ell^m(\cos\theta) \exp(im\phi) \tag{3.43}$$

with $q = \frac{1}{2}(n + \ell + 1)$.

In this section, we use QHJ formalism to obtain the solutions corresponding to the radial part of (3.41) which results from using separation of variables in the spherical coordinates. In these coordinates the Laplacian in (3.41) becomes

$$\nabla^2 = \frac{\partial^2}{\partial r^2} + \frac{2}{r}\frac{\partial}{\partial r} + \frac{1}{r^2}\left(\frac{\partial^2}{\partial \theta^2} + \cot\theta \frac{\partial}{\partial \theta} + \frac{1}{\sin^2\theta}\frac{\partial^2}{\partial \varphi^2}\right), \tag{3.44}$$

and the wave function can be separated into the radial and angular parts as

$$\Psi(r, \theta, \varphi) = \psi(r) f(\theta, \varphi). \tag{3.45}$$

For all separable problems in spherical coordinates, $f(\theta, \varphi) = Y_{\ell,m}(\theta, \varphi)$ is the eigenfunction of the θ and the φ eigenvalue equations. The radial equation of the three-dimensional oscillator is

$$-\frac{\partial^2 R(r)}{\partial r^2} + V_{\text{eff}}(r) R(r) = E R(r), \tag{3.46}$$

where $R(r) = \psi(r)/r$ and

$$V_{\text{eff}}(r) = \frac{1}{4}\omega^2 r^2 + \frac{\ell(\ell+1)}{r^2}. \tag{3.47}$$

For convenience, we subtract off the ground state energy and set up the QHJ formalism for the following potential

$$V^-(r) \equiv V_{\text{eff}}(r) = \frac{1}{4}\omega^2 r^2 + \frac{\ell(\ell+1)}{r^2} - \omega\left(\ell + \frac{1}{2}\right). \tag{3.48}$$

The eigenfunctions and eigenvalues for the above potential are obtained in the further analysis in the following sections.

QHJ analysis of the radial oscillator

The QHJ equation with the radial oscillator potential (3.48) is

$$q^2(r) + q'(r) + E - \frac{1}{4}\omega^2 r^2 - \frac{\ell(\ell+1)}{r^2} + \omega\left(\ell + \frac{3}{2}\right) = 0 \tag{3.49}$$

where $q(r) = (d \log R(r)/dr)$ and where the variable r is now to be treated as a complex variable. The QMF has a fixed pole at $r = 0$ and the point at infinity is assumed to be an isolated singularity. In addition, the QMF has n moving poles on the positive real line. It is to be pointed out that the real variable r takes values between 0 and ∞. In the QHJ formalism, since we work in the complex plane and due to the reflection symmetry, $r \to -r$, there will additional n moving poles on the negative real line, thus bringing the total number of moving poles to $2n$. We will assume that there are no other moving poles in the complex plane.

Form of QMF: The knowledge about the singularities allows us to write the QMF in the meromorphic form

$$q(r) = \frac{b_1}{r} + d_1 r + \sum_{k=0}^{2n} \frac{1}{r - r_k} + C. \tag{3.50}$$

Here, b_1 denotes the residue at $r = 0$ and d_1 represents the large r behavior of the QMF. The summation term represents the contribution of the $2n$ moving poles to the meromorphic form of the QMF. The residue at the moving poles is unity. By writing $\sum_{k=0}^{2n} \frac{1}{r - r_k} = \frac{P'_{2n}(r)}{P_{2n}(r)}$, where $P_{2n}(r)$ is a polynomial of degree $2n$, we get

$$q(r) = \frac{b_1}{r} + d_1 r + \frac{P'_{2n}(r)}{P_{2n}(r)} + C. \tag{3.51}$$

The constant C can be fixed from further analysis of QHJ equation as shown below.

Solution for QMF

Residue at $r = 0$: To calculate the residue at $r = 0$, we expand $q(r)$ in a Laurent expansion around $r = 0$ as

$$q = \frac{b_1}{r} + a_0 + a_1 r + \cdots \tag{3.52}$$

Substituting (3.52) in the QHJ equation (3.49) and comparing the coefficients of $1/r^2$, we obtain the following quadratic equation

$$b_1^2 - b_1 - \ell(\ell + 1) = 0 \tag{3.53}$$

which gives the two values

$$b_1 = \ell + 1; \quad b_1 = -\ell. \tag{3.54}$$

Calculation of d_1: The unknown constant d_1 can be fixed by following the steps used in Chap. 2 Sect. 2.4.2 for the harmonic oscillator. To calculate the large r behavior of the QMF, we perform a change the variable $r = 1/\zeta$ and the QHJ equation in terms of ζ is

$$q^2(\zeta) - \zeta^2 \frac{dq(\zeta)}{d\zeta} + E - \frac{\omega^2}{4\zeta^2} - \ell(\ell + 1)\zeta^2 - \omega \left(\ell + \frac{3}{2} \right) = 0. \tag{3.55}$$

The function $q(\zeta)$ is now expanded in a Laurent expansion around $\zeta = 0$. Since the most singular term in the above equation is $\omega^2/4\zeta^2$, the Laurent expansion will only have ζ^{-1} term with the coefficients of all higher-order negative powers of ζ becoming zero. Thus the Laurent expansion around $\zeta = 0$ becomes

$$q(\zeta) = \frac{d_1}{\zeta} + d_0 + d_2\zeta + \cdots . \tag{3.56}$$

Substituting this equation in (3.55) and comparing the coefficients of $1/\zeta^2$, we obtain the quadratic equation

$$d_1^2 - \frac{\omega^2}{4} = 0, \tag{3.57}$$

whose twin roots give the values of the residue d_1 as

$$d_1 = \frac{\omega}{2}; \quad d_1 = -\frac{\omega}{2}. \tag{3.58}$$

Choice of residues
In order to choose the right value of the constants b_1 and d_1, we use the boundary condition

$$q(r)\big|_{n=0} = -W(r), \tag{3.59}$$

where the function $W(r)$ is defined as

$$W(r) = -\frac{\psi_0'^{-}(r)}{\psi_0^{-}(r)} \tag{3.60}$$

with $\psi_0^{-}(r)$ being the ground state wave function of $V^{-}(r)$. For the radial oscillator potential, (3.48), the expression for $W(r)$ is

$$W(r) = \frac{\omega r}{2} - \frac{l+1}{r}. \tag{3.61}$$

In the literature, the function $W(r)$ is referred to as the superpotential and for more details we refer the reader to [7, 9] and more in Chap. 5. From the boundary condition (3.59), the QMF sans the moving part should be equal to the function $W(r)$. This leads to the following choice for the constants b_1 and d_1,

$$b_1 = \ell + 1 \quad \text{and} \quad d_1 = -\frac{\omega}{2}. \tag{3.62}$$

The meromorphic form of the QMF for the radial oscillator potential now becomes

$$q(r) = \frac{\ell+1}{r} - \frac{\omega r}{2} + \frac{P_{2n}'(r)}{P_{2n}(r)} + C. \tag{3.63}$$

Substituting this in the QHJ equation (3.49), we obtain the second-order differential equation

$$\frac{P_{2n}''(r)}{P_{2n}(r)} + 2\left[\frac{\ell+1}{r} - \frac{\omega r}{2} + C\right]\frac{P_{2n}'(r)}{P_{2n}(r)} + 2C\left[\frac{\ell+1}{r} - \frac{\omega r}{2}\right] + E + C^2 = 0.$$
(3.64)

Energy eigenvalues

In order to fix the value of C and obtain the energy eigenvalues, we look at the large r behavior of all the terms in the above equation. In the large r limit, *the leading behaviour is ωr* and equating the coefficients of r to zero, we get $C = 0$. We can obtain the expression for the wave function using the relation between $q(r)$ and the wave function $\psi(r) = \exp\left(\int q(r)dr\right)$.

Substituting $q(r)$ from (3.63), we obtain

$$\psi(r) = r^{\ell+1}\exp\left(-\frac{1}{4}\omega r^2\right)L_m^\alpha\left(\frac{1}{2}\omega r^2\right),$$
(3.65)

which is the expression for the eigenfunctions of the radial oscillator potential (3.48) apart from a normalization constant.

It is interesting to note that for all the ES potentials, $P_n(x)$ turns out to be one of the classical orthogonal polynomials [10]. For example, the harmonic oscillator has solutions in terms of the Hermite polynomials, while the radial oscillator and the hydrogen atom [8] potentials have solutions in terms of the Laguerre polynomials. Finally, the Jacobi polynomials appear in the solutions of the Scarf potential.

For our use in Chap. 5, we define $W_i(x) \equiv q(r)\big|_{n=0}$, $i = 1, 2, 3, 4$ corresponding to the four combinations of b_1 and d_1 as given in Table 3.1 .

$W_1(r)$, the superpotential for the case $i = 1$, equals the superpotential $W(r)$ given in equation Eq.(3.61) and led to the normalizable solution. The $W_i, i = 2, 3, 4$, lead to non-normalizable solutions, and these play a crucial role in the construction of rational potentials which have solutions in terms of exceptional orthogonal polynomials (EOPs) which are discussed in Chap. 5.

Table 3.1 Different values of b_1 and d_1 and the $W_i(r)$ describing the QMF at the fixed poles and the behaviour at infinity

S. No.	b_1	d_1	$W_i(r)$
1	$\ell+1$	$-\frac{\omega}{2}$	$-\frac{\ell+1}{r} + \frac{\omega}{2}$
2	$-\ell$	$-\frac{\omega}{2}$	$-\frac{\ell}{r} - \frac{\omega}{2}$
3	$\ell+1$	$\frac{\omega}{2}$	$\frac{\ell+1}{r} + \frac{\omega}{2}$
4	$-\ell$	$-\frac{\omega}{2}$	$-\frac{\ell}{r} - \frac{\omega}{2}$

3.5 Particle in a Box

One of the most common textbook problems is that of an infinite potential well. In this section we show how QHJ formalism can be used to solve this problem and find the energy eigenvalues. Many quantum mechanical problems require a change of variable from a real to a new real variable. *It will be noted from the following that in the QHJ treatment the change of variable for this problem is a mapping from a complex plane to another complex plane. This change of variable cannot be done in the Schrodinger wave mechanics. The QHJ formalism provides a natural framework to use a mapping from x to a new complex variable. The LPQEC, being a contour integral in the complex x- plane, can easily be written down as a contour integral in the new variable.*

We take the potential to be given by

$$V(x) = \begin{cases} 0, & \text{for } 0 < x < L, \\ \infty, & \text{for } x \le 0, \quad \text{or} \quad x \ge L. \end{cases} \quad (3.66)$$

The QMF obeys the QHJ equation:

$$p^2(x, E) - i\hbar \frac{\partial p(x, E)}{\partial x} = 2mE \qquad 0 \le x \le L. \quad (3.67)$$

The nodes of the eigenfunctions, and hence the moving poles of $p(x, E)$, are located between $x = 0$ and L. This is because the wave function vanishes identically outside this interval.

Let C be a thin rectangular contour enclosing the classically accessible region as shown in Fig. 3.1.

The LPEQC is given by

$$J(E) = \frac{1}{2\pi} \oint_c p(x, E) dx = n\hbar, \quad (3.68)$$

The contour encloses n moving poles each having residue $-i\hbar$, the quantization condition is, therefore, a straightforward application of the Cauchy residue theorem.

Fig. 3.1 Contour for $J(E)$

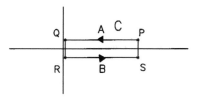

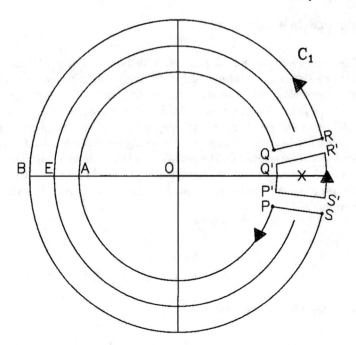

Fig. 3.2 Figure for particle in a box

To obtain the energy eigenvalues, we need to evaluate the contour integral Eq. (3.68) in terms of singularities outside the contour *PQRS*.

This is the simplest one-dimensional example where the motion of particle is restricted to the range $(0, L)$. This makes the problem unique and the method used for the harmonic oscillator in Chap. 1, is not applicable directly. A mapping of the complex x- plane described below solves the problem.

We use a mapping $z = \exp[(2\pi i x)/L]$; the contour C in the x-plane Fig. 3.2 is mapped onto the contour *PAQRBSP* in the z-plane. The moving poles get mapped on the middle arc of unit radius, see Fig. 3.2. The quantization condition, in terms of the new variable z, takes the form

$$J(E) = \frac{L}{4i\pi^2} \oint_{C_1} \frac{\tilde{p}(z, E)\mathrm{d}z}{z} = n\hbar, \tag{3.69}$$

where C_1 is the contour *PAQRBSP* of Fig. 3.2.

The QHJ equation, written in terms of the new variable z, is

$$\tilde{p}^2(z, E) + \frac{2\pi \hbar z}{L} \frac{\partial \tilde{p}(z, E)}{\partial z} = 2mE, \tag{3.70}$$

where $\tilde{p}(z, E) = p\left(L \log(z)/(2\pi i), E\right)$.

The boundary condition that the wave function vanishes at $x = 0$ and $x = L$ gives rise to a pole in $\tilde{p}(z, E)$ at $z = 1$. Let γ and Γ be the inner and outer *full* circles of radii OA and OB, respectively, both taken in the counterclockwise direction. The integral in (3.69) can be written in terms of integrals over Γ, γ, and the contour $P'S'R'Q'P'$ enclosing the pole at $z = 1$. Thus we get

$$J(E) = \frac{L}{4i\pi^2} \left(\oint_\Gamma \frac{\tilde{p}(z, E)dz}{z} - \oint_\gamma \frac{\tilde{p}(z, E)dz}{z} - \oint_{P'S'R'Q'P'} \frac{\tilde{p}(z, E)dz}{z} \right) \quad (3.71)$$

The first integral is computed by changing variables from z to $y = 1/z$, as was done in Sect.2.5 for the harmonic oscillator. The last two integrals in the above expressions are calculated as usual by computing the residues of $\tilde{p}(z, E)$ at $z = 0$ and $z = 1$, and we have

$$\oint_\Gamma \frac{\tilde{p}(z, E)dz}{z} = -2\pi i \sqrt{2mE}, \quad (3.72)$$

$$\oint_\gamma \frac{\tilde{p}(z, E)dz}{z} = 2\pi i \sqrt{2mE}, \quad (3.73)$$

$$\oint_{P'S'R'Q'P'} \frac{\tilde{p}(z, E)dz}{z} = \frac{4\pi^2 i \hbar}{L}. \quad (3.74)$$

Substituting the above in (3.71), and solving for E we get

$$E = \left(\frac{\hbar^2}{2mL^2} \right) (n + 1)^2, \qquad n = 0, 1, 2, \ldots . \quad (3.75)$$

3.6   Exactness of SWKB Approximation

The supersymmetric WKB approximation is similar to the WKB approximation, and is known to give exact eigenvalues for certain potentials [13]. It has been extensively studied in literature.

Assuming that the potential supports classical bounded motion with two turning points, x_1, x_2, the standard lowest order WKB approximation formula for the energy eigenvalues is

$$J_{\text{WKB}} \equiv \frac{1}{\pi} \int_{x_1}^{x_2} \sqrt{2m(E - V(x))}dx = (n + 1/2)\hbar. \quad (3.76)$$

The SWKB approximation consists in first writing the potential as

$$V(x) = W^2(x) - W'(x), \tag{3.77}$$

where $W(x)$ is the logarithmic derivative of the quantum mechanical ground state wave function.

The SWKB approximation consists in using

$$J_{\mathrm{SWKB}} \equiv \frac{1}{\pi} \int_{x_1}^{x_2} \sqrt{2m(E - W^2)}\mathrm{d}x = n\hbar, \tag{3.78}$$

to determine the energy eigenvalues. Note that only W^2, and not the full potential $V(x)$, appears inside the integral. Also the right-hand side of Eq. (3.78) has $n\hbar$ as against $(n + 1/2)\hbar$ in the WKB approximation Eq. (3.76). While it is known that the SWKB approximation gives exact results for all translationally shape invariant potentials listed in [14, 15], numerical studies show that there exist potential exceptions for which SWKB is neither exact, nor it is better that the standard WKB approximation [16]. This raises the question when SWKB is exact and when it is not. In this connection, we mention that for a class of potentials, corrections to SWKB approximation has been have been investigated. Raghunathan et al. [17] write the integral J_{SWKB} as a contour integral in complex plane and show that corrections of higher orders in powers of $\hbar$ in the eigenvalue expressions vanish.

Introducing the notation $Q_{\mathrm{SWKB}}(x) \equiv \sqrt{2m(E - W^2)}$ for convenience, the SWKB rule of Eq. (3.78) becomes

$$J_{\mathrm{SWKB}} = \frac{1}{\pi} \int_{x_1}^{x_2} Q_{\mathrm{SWKB}}(x)\,\mathrm{d}x = n\hbar. \tag{3.79}$$

Here we will explain only the main idea how LPEQC can be used to determine if SWKB gives exact results or not. As a first step, the function $Q_{\mathrm{QWKB}}(x) \equiv \sqrt{2m(E - W^2)}$ is defined as a multi valued function of a complex variable x. We follow [17] and rewrite the J_{SWKB} as an integral in the complex plane around the branch cut from x_1 to x_2. The details of the steps involved are similar to an example given in Sect. 6.3 of Part-I of [18]. Thus we arrive at the SWKB condition in the following form

$$J_{\Gamma} = \frac{1}{2\pi i} \oint_{\Gamma} \sqrt{2m(E - W^2(x))}\mathrm{d}x n\hbar. \tag{3.80}$$

Here Γ is an anticlockwise contour enclosing the branch cut from x_1 to x_2.

We recall the LPEQC in terms of the action integral is

$$J(E) = \frac{1}{2\pi i} \oint_C p(x)\mathrm{d}x = n\hbar \tag{3.81}$$

where $p(x)$ is the QMF for the given potential $V(x)$.

In cases of interest, where SWKB is found to be exact, the contours Γ can be deformed into C without crossing a singular point.

The contour integrals in (3.81) and (3.80) can be computed in terms of the residues at the singularities outside the respective contours. Since LPEQC gives exact energy eigenvalues, comparing Eq. (3.80) with (3.81), we see that SWKB rule will be exact if the pole structure and the corresponding residues of both the integrands match.

This is exactly what happens for all the potentials for which the SWKB approximation gives exact answers for the eigenvalues. These statements about the locations of poles and the residues of Q_{SWKB} are easily verified in explicit examples. Taking the example of radial oscillator, with $W(r)$ given by Eq. (3.61), is obvious that the function Q_{SWKB}

$$Q_{\mathrm{SWKB}} = \sqrt{2m\left(E - W^2(r)\right)} \tag{3.82}$$

has poles at $r = 0$ and at infinity. It is not difficult to verify that the corresponding residues at these poles coincide with those for the QMF already found in the Sect. 3.3. It then follows that the integral J_{SWKB} equals the action integral $J(E)$ and therefore the SWKB will give the exact eigenvalues. Full details on exactness of SWKB approximation, within the QHJ approach, for several potentials including the rational potentials discussed in Chap. 5 can be found in [19, 20]. Finally, we wish to point out that that approximate wave functions

$$\psi(x) \sim \frac{1}{\sqrt{Q_{\mathrm{SWKB}}}} \exp\left(\pm \frac{i}{\hbar} \int Q_{\mathrm{SWKB}}(x)\,\mathrm{d}x\right) \tag{3.83}$$

will not be exact eigenfunctions.

3.7 Concluding Remarks

Several other ES models have been investigated within the QHJ framework. For example, see [3, 4, 11] for detailed QHJ treatment of of potential models with shape invariance and see [12] for hydrogen atom eigenvalues and eigenfunctions.

A few important observations will be made about the ES models that have been studied in articles referred above. The QMF turns out to be a rational function in the complex plane after a suitable change of the independent variable has been made. In almost all the cases, the change of variable has been easy to guess and it transformed the QHJ equation into another equation of similar form with rational coefficients. For all the potentials studied the point at infinity turned out to be an isolated singular point. It is this property that makes it possible to compute the integral appearing in LPEQC.

Next, the full expression of QMF and hence the eigenfunctions gets determined by a straightforward application of well-known Liouville theorem on meromorphic functions. In the next chapter we will be discussing some potentials with exotic properties. Solution of these potentials required new inputs, and investigation led to new and unexpected results; new in the sense that these could not have been anticipated on the basis of study of ES models.

References

1. Leacock, R.A., Padgett, J.: Hamilton-Jacobi theory and the quantum action variable. Phys. Rev. Lett. **50**, 3–6 (1983)
2. Leacock, R.A., Padgett, M.J.: Hamilton-Jacobi action angle quantum mechanics. Phys. Rev. D **28**, 2491–2502 (1983)
3. Bhalla, R.S., et al.: Quantum Hamilton-Jacobi formalism and bound state spectra. Am. J. Phys. **65**, 1187–1193 (1997)
4. Bhalla, R.S.: The Quantum Hamilton-Jacobi formalism approach to energy spectra of potential problems, Ph. D. Thesis submitted to University of Hyderabad (1996)
5. Landau, L.D., Lifshitz, E.M.: Quantum Mechanics, 3rd edn. Non relativistic Theory. Pergamon Press, Oxford (1977)
6. Hille, Einar: Ordinary Differential Equations in Complex Domain. Wiley Interscience Publications, New York (1976)
7. Cooper, F., et al.: Supersymmetry in quantum mechanics. Phys. Rep. **251**, 267–385 (1995)
8. Mathews, P.M., Venkateshan, K.: A Textbook of Quantum Mechanics. Tata McGraw-Hill Publishing Co., New Delhi (1976)
9. Cooper, F., et al.: Supersymmetric Quantum Mechanics. World Scientific, Singapore (2001)
10. Dennery, Philippe, Kryzwicki, André: Mathematics for Physicists. Dover Publications, New York (2012)
11. Sree Ranjani S. et al.: Bound state wave functions through QHJ formalism. Mod. Phys. Lett. A **19**, 1457–1468 (2004)
12. Sree Ranjani, S.: Quantum Hamilton-Jacobi solution for spectra of several one dimensional potentials with special properties, Ph. D. Thesis submitted to University of Hyderabad (2004)
13. Comtet, et al.: Phys. Lett. B **150**, 159 (1985)
14. Cooper, F., Khare, A., Sukhatme, U.P.: Supersymmetry and quantum mechanics. Phys. Rep. **251**, 267–385 (1995)
15. Cooper, F., Khare, A., Sukhatme, U.P.: Supersymmetric quantum mechanics. World Scientific Publishing Co. Ltd. Singapore (2001)
16. DeLenay, D. et al.: Phys. Lett. B **247**, 301 (1990)
17. Raghunathan, K., et al.: Phys. Lett. B **188**, 351 (1987)
18. Kapoor, A.K.: Complex Variables—Principles and Problem Sessions. World Scientific Publishing Co., Singapore (2011)
19. Bhalla, R.S., et al.: Exactness of the supersymmetric WKB approximation scheme. Phys. Rev. A **54**, 951 (1996)
20. Sree Ranjani, S., et al.: The exceptional orthogonal polynomials, QHJ formalism and the SWKB quantization condition. J. Phys. A Math. Theor. **45**, 055210 (2012)

Chapter 4
Exotic Potentials

4.1 Introduction

The potential models discussed in the previous chapter were all ES models either
in one dimension or models arising out of separation of variables in more than one
dimension. Application of the QHJ formalism to these and similar models is straight
forward and did not offer any new difficulty.

- A first step in almost all our studies has been a change of variable so that the QHJ
  equation has the form of Riccati equation with rational coefficients.
- The number of moving poles of QMF turns out to be finite in the new variable.
- For models in one dimension, $-\infty < x < \infty$, the moving poles appear only in the
  classical region. The location and the number of moving poles are in a one to one
  correspondence with the nodes of the wave function as predicted by oscillation
  theorems in the theory of differential equations.
- The point at infinity turns out to be an isolated singularity of the QMF. In all the
  potential models that we have studied, this property is found to hold after a suitable
  change of variable has been made.
- The QMF does not have any other moving pole, or any other singularity in the
  complex plane.
- The point at infinity, being an isolated singularity appears to be an important
  property of the QMF of all ES models. In all the cases that we have studied using
  the QHJ formalism, it has led to an exact analytical solution in a finite number of
  steps. Thus it can be a starting point for defining an ES potential model in quantum
  mechanics.

In this chapter we study a few models with *exotic* properties. For these models the
QHJ scheme does not proceed in a routine manner, as was the case for ES models;
they present new challenges that need to be addressed before QHJ formalism can
be applied. We present a brief summary of some of these aspects in this section.
More details will be found in the corresponding sections of this chapter, and related

© The Author(s), under exclusive license to Springer Nature Switzerland AG 2022
A. K. Kapoor et al., *Quantum Hamilton-Jacobi Formalism*,
SpringerBriefs in Physics, https://doi.org/10.1007/978-3-031-10624-8_4

references can be found in the two doctoral theses [1, 2] submitted to the University of Hyderabad.

- The Scarf potential[1] [4] supports both energy bands and bound states for different values of potential strengths. A natural question that needs to be investigated is, *"How do both types of spectra arise from a single Leacock-Padgett quantization condition?"*
- The Scarf-I potential [1] exhibits two phases of supersymmetry [3]. This means that for different ranges of potential parameters, the bound state energy spectrum is different.*Again it is not clear how QHJ formalism can handle such a situation?*
- The QES models pose a different kind of challenge. For these models, exact analytical solution, for the energy levels and eigenfunctions, is known to exist [5] for *only a part of spectrum*. Even for this part of the spectrum, the solutions are known only if the potential parameters obey a condition known as the quasi-exact solvability condition (QES condition). The QES models have been extensively studied using group theoretic, algebraic and other methods. The simplest QES model is a sextic oscillator with potential

$$V(x) = \alpha x^2 + \beta x^4 + \gamma x^6, \tag{4.1}$$

and the corresponding QES condition is

$$\alpha = \frac{\beta^2}{4\gamma} - (2\nu + 3)\sqrt{\gamma}. \tag{4.2}$$

where ν is a non-negative integer. The challenge posed by the QES models is to understand,*"How does the QES condition arise and how only a subset of the energy levels can be found within the QHJ scheme?"*

In order to attack the above mentioned problems using the QHJ formalism, one needs some idea about the singularity structure of the QMF. Some possible clues can be obtained by looking at $\hbar \to 0$ limit. In this limit the QMF should approach the classical momentum function. Taking a simple example of the particle in a box, Leacock and Padgett have discussed how moving poles can condense and give rise to branch cut in the classical momentum function in the limit of $\hbar \to 0$. The classical momentum function for the sextic oscillator is

$$p_{cl} = \sqrt{(E - \alpha x^2 - \beta x^4 - \gamma x^6)} \tag{4.3}$$

and will, in general, have six branch points in the complex plane. The location of branch points and branch cuts depends on the energy and the parameters α, β, γ. This line of thinking does not offer any clue about the singularity structure of the QMF in the complex plane.

We take up the details of the strategy adopted to address this issue in Sect. 4.4. We could solve the problem and reproduce all the results already known in the

---

[1] This coincides with the potential listed as Scarf-I in [3] with $B = 0$.

literature for different QES models [5]. This study led to new information about the zeros of the wave function.

- The Lamé periodic potential discussed in Sect. 4.5 is defined in terms of the elliptic functions. A suitable change of variable was required to arrive at the QHJ equation with rational potential. Once a change of variable [1] was found, it was straightforward to obtain the solutions for the band edges. More details about this point are given in Sect. 4.5.

4.2 A Periodic Potential

We take the Scarf potential [4] as

$$V(x) = -V_0 \cosec^2 \pi(x/a). \tag{4.4}$$

In Figs. 4.1 and 4.2, the potential is plotted, respectively, for $V_0 < 0$ and $V_0 > 0$. If $V_0 < 0$, it is seen that the potential becomes infinite as $x \to n\pi$, n — integer. Therefore the particle motion is confined to one period only and the allowed quantum states are bound states only. For $V_0 > 0$ and entire real line is the physically accessible region. Like any periodic potential, we have has band spectrum. We will be interested only in solutions for energies and eigenfunctions for the band edges.

We shall write V_0 as

$$V_0 = \frac{\hbar^2 \pi^2}{2ma^2}\left(\frac{1}{4} - s^2\right) \tag{4.5}$$

and henceforth work with the parameter s. The treatment given here closely follows [1], where more details can be found. We define parameters λ and s by

$$\lambda^2 = \frac{2mEa^2}{\pi^2\hbar^2}, \qquad \left(\frac{1}{4} - s^2\right) = \frac{2mV_0a^2}{\pi^2\hbar^2} \tag{4.6}$$

Fig. 4.1 $V_0 = -1, a = 1$

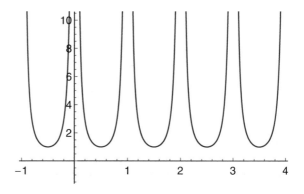

Fig. 4.2 $V_0 = 1, a = 1$

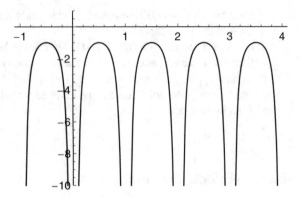

and we will write most of our equations in terms of these two parameters. The Schrodinger equation takes the form

$$\frac{d^2\psi}{dx^2} + \frac{a^2}{\pi^2}\left(\lambda^2 + \frac{(1/4 - s^2)}{\sin^2(\pi x/a)}\right)\psi = 0. \tag{4.7}$$

Changing variable to $y = \cot(\pi x/a)$ and following the steps given in Sect. 3.2, the QHJ equation takes the form

$$\chi^2 + \frac{d\chi}{dy} + \frac{(\lambda^2 - 1)}{(y^2 + 1)^2} + \frac{\frac{1}{4} - s^2}{(y^2 + 1)} = 0, \tag{4.8}$$

where

$$\chi = \phi + \frac{y}{(1 + y^2)}, \quad \phi(y) = \frac{aq(x_1, E)}{\pi(1 + y^2)} \tag{4.9}$$

and $q(x, E) = ip(x, E)$. Equation (4.8) shows that χ has fixed poles at $y = \pm i$ and the singular part of χ can be written as

$$\chi_{\text{sing.}} = \frac{b_1}{y - i} + \frac{b_1'}{y + i}. \tag{4.10}$$

If y_k are locations of the moving poles then χ can be written as

$$\chi = \chi_{\text{sing.}} + \sum_k \frac{1}{y - y_k} + \Phi(y). \tag{4.11}$$

Here we have used a result from Sect. 2.4.1 that the residue at a moving pole is unity. As $y \to \infty$, $\chi \to 0$ and $\Phi(y)$ is holomorphic and bounded. Therefore, by Liouville's theorem $\Phi(y)$ must be a constant, say C. Thus we get

$$\chi = \chi_{\text{sing.}} + \sum_{y_k} \frac{1}{y - y_k} + C. \tag{4.12}$$

The residues b_1, b_1' are computed as usual by substituting the Laurent expansion in powers of $(y \pm i)$ in the QHJ equation and equating coefficients of different powers of $(y \pm i)$ to zero. This gives

$$b_1 = \frac{1 \pm \lambda}{2}, \quad b_1' = \frac{1 \pm \lambda}{2}. \tag{4.13}$$

The points $y = \pm i$ correspond to complex values of x. Therefore, requirement that the eigenfunction $\psi(x)$ be square integrable does not give rise to any restriction on the values of residues b_1 and b_1'. We must find some other criteria to choose the allowed values of b_1 and b_1'. It should be noted that the Scarf potential is an even function of x and hence the eigenfunctions can be chosen to have definite parity. Thus $\psi(\pm x) = \pm \psi(x)$ and $q(x) = \frac{1}{\psi(x)} \frac{d}{dx} \psi(x)$ must be an odd function of x and hence an odd function of y. This gives the constraint $\chi(-y) = -\chi(y)$ and therefore

$$b_1 = b_1'. \tag{4.14}$$

The potential becomes infinite at $x = \pm a$ and we need to check the behaviour of the solution $\psi(x) = \exp(\int q(x)dx)$ near $x = \pm a$. These points correspond to $y = \pm \infty$, and one can obtain the behaviour of $\chi(y)$ at infinity by changing variable to $t = 1/y$ and expanding $\chi(y)$ in powers of t.

$$\chi = a_0 + \frac{d_1}{t} + \frac{d_2}{t^2} + \cdots . \tag{4.15}$$

Substituting in the QHJ equation, written in terms of variable t, leads to

$$d_1^2 - d_1 + \left(\frac{1}{4} - s^2\right) = 0, \quad \Rightarrow d_1 = \frac{1 \pm 2s}{2}. \tag{4.16}$$

Note that $-d_1$ is negative of the residue of $\chi(y)$ at $y = \infty$ and will be needed later. We shall assume that the number of moving poles is finite and that χ can be written as

$$\chi = \frac{b_1}{y - i} + \frac{b_1'}{y + i} + \frac{P'(y)}{P(y)} + C, \tag{4.17}$$

where $P(y) = \prod_{k=1}^{n}(y - y_k)$ is a polynomial of degree n. Thus χ is a rational function of y and the sum of residues at all poles, including that at infinity vanishes. This requirement imposes the following restriction

$$b_1 + b_1' + n = d_1. \tag{4.18}$$

Case I: Bound states

This case corresponds to $(s > \frac{1}{2})$. For bound states $d_1 = \frac{1}{2} - s < 0$ and out of the two possible values of b_1 and b_1', only the values $b_1 = b_1' = 1 - \frac{\lambda}{2}$ is acceptable. The other value $b_1 = b_1' = 1 + \frac{\lambda}{2}$ violates the condition $d_1 - 2b_1 \geq 0$. Recalling $\lambda^2 = 2mEa^2/\pi^2$, we get the energy eigenvalues as

$$E_n = \frac{\pi^2}{2ma^2} \left(\frac{1}{2} + n + \sqrt{\frac{1}{4} - \frac{2mV_0a^2}{\pi^2\hbar^2}} \right)^2. \tag{4.19}$$

Explicit expressions for wave functions are obtained using the form (4.17),

$$\psi(y) = (1 + y^2)^{b_1} \frac{P'(y)}{P(y)}, \tag{4.20}$$

where $P(y)$ is a polynomial of degree n. Substituting χ from (4.17) in QHJ equation, we get a second-order differential equation which turns out to be the equation for a Jacobi polynomial. Therefore, the final answer for the bound state wave functions is easily found to be

$$\psi(y) = (y^2 + 1)^{-\lambda/2} P_n^{s_1, s_2}(-iy), \qquad s > \frac{1}{2}, \tag{4.21}$$

where $s_1 = s_2 = -n - s - \frac{1}{2}$.

Case II Band spectrum

We now consider the case $0 < s < \frac{1}{2}$ for which the Scarf potential has energy bands. We shall determine the eigenvalues and eigenfunctions for the band edges. Using the requirement that $d_1 - b_1 - b_1' = d_1 - 2b_1 \geq 0$, we see that only following combinations are allowed $(\lambda > 0)$

$$\text{(i) } b_1 = b_1' = \frac{1 - \lambda}{2}, \quad d_1 = \frac{1}{2} - s, \quad \lambda = (s + \frac{1}{2}) + n, \tag{4.22}$$

$$\text{(ii) } b_1 = b_1' = \frac{1 - \lambda}{2}, \quad d_1 = \frac{1}{2} + s, \quad \lambda = -(s - \frac{1}{2}) + n. \tag{4.23}$$

The two sets correspond to energy values

$$E_n^{\pm} = \frac{\pi^2}{2ma^2} \left(n + \frac{1}{2} \pm s \right)^2 \tag{4.24}$$

E_n^+ and E_n^- corresponds to the upper and lower band edges, respectively. The wave functions for the band edges are easily seen to be given by

$$\psi(y) = (y^2 + 1)^{-\lambda/2} P_n^{v_1, v_2}(-iy), \tag{4.25}$$

where $v_1 = n + 2 + \frac{1}{2}$, $v_2 = n - s + \frac{1}{2}$.

These results are in agreement with those in [4]. It may be remarked that the wave functions for energies inside the bands have not been obtained. One needs to have an idea about the singularity structure and form of the QMF to make any progress in this direction within the QHJ formalism.

4.3 A Potential with Two Phases of SUSY

4.3.1 QHJ Solution of Scarf-I Potential

The expression for the supersymmetric Scarf-I potential, see Table 4.1 in [3], is,

$$V_-(x) = -A^2 + (A^2 + B^2 - A\alpha\hbar)\sec^2 \alpha x - B(2A - \alpha\hbar)\tan \alpha x \sec \alpha x. \quad (4.26)$$

We need to retain $\hbar$ and do not set it equal to one. This is because, in order to make use of LPBC, we need the leading behaviour of the QMF in the limit $\hbar \to 0$. Note that the potential becomes infinite as $x \to \pm\pi/2$. The turning points will lie in the range $(-\pi/2, \pi/2)$. This problem exhibits broken and unbroken phases of SUSY. The spectra and the energy eigenfunctions in the two phases are different.

In the parameter range,

$$(A - B) > 0, \quad (A + B) > 0, \quad (4.27)$$

SUSY is exact and SUSY is broken for the range

$$(A - B) > 0, \quad (A + B) < 0. \quad (4.28)$$

This potential poses a question, "How can two different sets of solutions arise in the QHJ formalism?" The supersymmetric phases of this potential are well studied and their corresponding bound state solutions are available in the literature. In this section, we will show that the ranges of the two different phases and their corresponding solutions emerge naturally from a careful application of the LPBC.

Setting $2m = 1$, but retaining $\hbar$ and writing the QHJ equation in terms of $q = ip = \hbar\frac{d\log\psi(x)}{dx}$, we get the QHJ equation

$$q^2 + \hbar q' + E + A^2 - (A^2 + B^2 - A\alpha\hbar)\sec^2 \alpha x + B(2A - \alpha\hbar)\tan \alpha x \sec \alpha x = 0. \quad (4.29)$$

4.3.2   Change of Variable

We perform a change of variable

$$y = \sin \alpha x, \tag{4.30}$$

Using notation of Ch2 §5.2 we have $F(y) = \alpha\sqrt{1 - y^2}$ and the transformation equations take the form

$$q = \alpha\sqrt{1 - y^2}\phi; \quad \phi = \chi + \frac{y\hbar}{2(1 - y^2)}. \tag{4.31}$$

The QHJ equation in terms of χ is obtained as

$$\chi^2 + \hbar\chi' + \frac{y^2\hbar^2}{4(1 - y^2)^2} + \frac{E + A^2}{\alpha^2(1 - y^2)} + \frac{\alpha^2\hbar - 2(A^2 + B^2 - A\alpha\hbar)}{2\alpha^2(1 - y^2)^2}$$

$$+ \frac{B(2A - \alpha\hbar)y}{\alpha^2(1 - y^2)^2} = 0. \tag{4.32}$$

It is useful to have an expression for χ in terms of the QMF. Using the transformation equation (4.31), we have

$$\chi = \phi - \frac{y\hbar}{2(1 - y^2)} = \frac{q}{\alpha\sqrt{1 - y^2}} - \frac{y\hbar}{2(1 - y^2)} \tag{4.33}$$

$$= \frac{i\hbar p}{\alpha\sqrt{1 - y^2}} - \frac{y\hbar}{2(1 - y^2)}. \tag{4.34}$$

4.3.3   Meromorphic Form of the QMF χ

From (4.32), it is seen that χ has fixed poles at $y = \pm 1$. In addition $\chi(y)$ will have n moving poles, with residue one. Assuming that χ has only these singularities in the complex plane, we write χ in the meromorphic form as

$$\chi = \frac{b_1}{y - 1} + \frac{b_1'}{y + 1} + \hbar\frac{P_n'(y)}{P_n(y)} + C. \tag{4.35}$$

Here $P_n(y)$ is an nth degree polynomial. The constants b_1 and b_1' are the residues at $y = \pm 1$ respectively. Since $\chi(y)$ is bounded at infinity, a straightforward application of Liouville's theorem theorem shows that the analytic part of $\chi(y)$ is a constant C.

In the present case it turns out to be necessary to use the boundary condition in the form originally proposed by Leacock and Padgett. We first define the classical momentum function, which will help us to fix the right form of χ.

4.3.4 Computation of the Residues Using QHJ

We shall now compute the required residues using the QHJ equation.

Residue at $y = -1$:

The residue of χ at $y = -1$ will be evaluated by making use of the Laurent series expansion of each term in (4.32) in powers of $(y + 1)$.

For χ the Laurent expansion around $y = -1$ will be taken to be

$$\chi = \frac{b'_1}{y + 1} + a_0 + a_1(y + 1) + \cdots \cdots . \tag{4.36}$$

Comparing coefficients of different powers of $y + 1$ gives two values of b'_1 as follows.

$$b'_1 = \frac{(A + B)}{2\alpha} + \frac{\hbar}{4}, \qquad -\frac{(A + B)}{2\alpha} + \frac{3\hbar}{4}. \tag{4.37}$$

Residue at $y = 1$:

Repeating the above calculation, we get the two values of the residue of χ as

$$b_1 = \frac{(A - B)}{2\alpha} + \frac{\hbar}{4}, \qquad -\frac{(A - B)}{2\alpha} + \frac{3\hbar}{4}. \tag{4.38}$$

Residue at infinity

The QHJ equation for χ is

$$\chi^2 + \hbar\chi' + \frac{y^2\hbar^2}{4(1 - y^2)^2} + \frac{E + A^2}{\alpha^2(1 - y^2)} + \frac{\alpha^2\hbar - 2(A^2 + B^2 - A\alpha\hbar)}{2\alpha^2(1 - y^2)^2}$$
$$+ \frac{B(2A - \alpha\hbar)y}{\alpha^2(1 - y^2)^2} = 0. \tag{4.39}$$

To determine the behaviour of the QMF for large y we make a change of variable to $\zeta = 1/y$ and write the (4.39) in terms of the new variable ζ

$$\tilde{\chi}^2 - \zeta^2\hbar\tilde{\chi}' + \frac{\zeta^2\hbar^2}{4(1 - \zeta^2)^2} - \frac{(E + A^2)\zeta^2}{\alpha^2(1 - \zeta^2)} + \frac{(\alpha^2\hbar - 2(A^2 + B^2 - A\alpha\hbar)\zeta^4}{2\alpha^2(1 - \zeta^2)^2}$$
$$+ \frac{B(2A - \alpha\hbar)\zeta^3}{\alpha^2(1 - \zeta^2)^2} = 0, \tag{4.40}$$

where $\tilde{\chi}(\zeta) = \chi(1/\zeta)$. To obtain large y behaviour, we expand different terms in Taylor series in powers of ζ. For small ζ, the most dominant terms are the ζ^2 terms. Writing

$$\tilde{\chi} = c_0\zeta + O(\zeta^2) \tag{4.41}$$

we get an equation for c_0

$$c_0^2 - \hbar c_0 + \frac{\hbar^2}{4} - (E + A^2)/\alpha = 0 \tag{4.42}$$

and the two solutions for c_0 are

$$c_0 = \frac{\hbar}{2} \pm \frac{1}{\alpha}\sqrt{E + A^2}. \tag{4.43}$$

This gives us two answers for the residue of $\chi(y)$ at infinity, being equal to $-c_0$:

$$\text{Res}\{\chi(y)\}_\infty = -\frac{\hbar}{2} \pm \frac{1}{\alpha}\sqrt{E + A^2}. \tag{4.44}$$

Since the QHJ for χ is a quadratic equation, the use of this equation gives two answers for the residues. Therefore, we need to apply suitable boundary condition to select appropriate values of constants b_1, b_1'. For this purpose, we will implement Leacock-Padgett boundary condition to choose value corresponding to physically acceptable bound state wave functions.

4.3.5 *Leacock-Padgett Boundary Condition*

The aim of present section is to show that LPBC automatically selects the range of parameters for the two phases of SUSY. This in turn leads to answers in agreement with the literature for both the phases of SUSY. To use LPBC, first we need to define the classical momentum function as a single-valued function in the cut complex plane.

4.3.5.1   Classical Momentum

We begin with the expression

$$p_c = \sqrt{2m(E - V(x))} \tag{4.45}$$

This is a multivalued function and a careful definition is needed. We assume the branch cut to be between the two turning points x_1 and x_2 in the complex x- plane. Moreover a choice of the branch is made so that the QMF may be positive just below the branch cut.

The entire analysis will be performed after a change of variable from x to y. Therefore, we will now define $\chi(y)$ in the complex y- plane, see Eq. (4.52). The classical momentum function expressed in terms of y takes the form

Fig. 4.3 Polar variables to define p_c

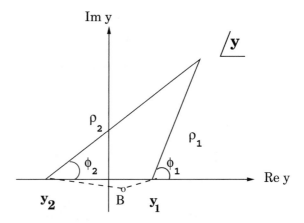

$$\tilde{p}_c = \pm i \frac{\sqrt{E + A^2}}{\sqrt{1 - y^2}} \left(y^2 + \frac{B^2 - A\alpha - E - B(2A - A\alpha\hbar)y}{E + A^2} \right)^{1/2}, \qquad (4.46)$$

We denote the two turning points, in the x- plane, as x_1, x_2. We note that the images of these two points y_1, y_2 in the y- plane lie in the interval $(-1, 1)$. We can therefore write p_c as

$$p_c = \pm i \frac{\sqrt{E + A^2}}{\sqrt{1 - y^2}} \sqrt{(y - y_1)(y - y_2)} \qquad (4.47)$$

This is a multivalued function in complex y plane and a proper definition be given before LPBC can be used.

Let y_1 and y_2 be the images of the two turning points x_1 and x_2 in the y plane. The mapping from $y = \sin \alpha x$, is such that it maps the point just below the branch cut in the x plane to a point just below the branch cut in the y plane. We introduce the variables ϕ_1, ρ_1 and ϕ_2, ρ_2, where

$$y - y_1 = \rho_1 \exp(i\phi_1), \qquad y - y_2 = \rho_2 \exp(i\phi_2) \qquad (4.48)$$

and the ranges of ϕ_1, ϕ_2 are chosen to be

$$\phi_1 \in (0, 2\pi), \qquad \phi_2 \in (0, 2\pi), \qquad (4.49)$$

with $(\rho_1 + \rho_2) > 2$. Our choice corresponds to a branch cut for the classical momentum function between y_2 and y_1, see Fig. 4.3.

Recall that LPBC demands that p_c be positive below the cut in x-plane. It is not difficult to check that p_c must be chosen to be positive for a point below the branch cut in the complex y plane also. For a point just below the branch cut we have $\phi_1 = \pi$, $\phi_2 = 2\pi$. Requiring LPBC in y plane, we see that the definition of the classical momentum function p_c in the complex y plane must be taken as

$$p_c = i \frac{\sqrt{E + A^2}}{\sqrt{1 - y^2}} \sqrt{(y - y_1)(y - y_2)} \tag{4.50}$$

$$= i \frac{\sqrt{E + A^2}}{\sqrt{1 - y^2}} (\rho_1 \rho_2)^{1/2} e^{i(\phi_1 + \phi_2)/2} \tag{4.51}$$

obtained by choosing the positive sign in (4.47). The corresponding equation for χ_c is now determined as

$$\chi_c = -\frac{1}{\alpha} \frac{\sqrt{E + A^2}}{1 - y^2} (\rho_1 \rho_2)^{1/2} e^{i(\phi_1 + \phi_2)/2} - \frac{y\hbar}{2(1 - y^2)} . \tag{4.52}$$

4.3.5.2   Choice of Residues of χ_c at the Fixed Poles

Residue at $y = -1$:

Starting from the expression of p_c, we first derive an expression for corresponding χ_c using χ, after making the replacement $p \to p_c$ we calculate the residue of χ_c at the fixed pole $y = -1$. In the complex plane, the point $y = -1$ corresponds to $\phi_1 \sim \phi_2 \sim \pi$, which gives

$$\chi_c = \frac{\sqrt{E + A^2}}{\alpha(1 - y^2)} (\rho_1 \rho_2)^{1/2} - \frac{y\hbar}{2(1 - y^2)} . \tag{4.53}$$

The residue of χ_c at $y = -1$ is

$$\sim \frac{1}{\alpha} \sqrt{\frac{E + A^2}{2}} (\rho_1 \rho_2)^{1/2} + \frac{\hbar}{4} , \tag{4.54}$$

which is a positive quantity.

Since the residue of χ_c at $y = -1$ is positive, (4.54) implies that we select the value of b'_1, which is positive in the limit $\hbar \to 0$ Therefore we must make the following choices for the residues of χ

$$b'_1 = \begin{cases} \frac{A + B}{2\alpha} + \frac{\hbar}{4} & \text{if } (A + B) > 0 \\ -\frac{A + B}{2\alpha} + \frac{3\hbar}{4} & \text{if } (A + B) < 0 \end{cases} . \tag{4.55}$$

Residues at $y = 1$:

Repeating the above process at the other fixed pole, we obtain the residue of χ_c at $y = 1$ as

$$\sim \frac{1}{\alpha} \frac{\sqrt{E + A^2}}{2} (\rho_1 \rho_2)^{1/2} + \frac{\hbar}{4} , \tag{4.56}$$

which is positive. Comparing Eqs.(4.56) and (4.38), we see that each value of b_1 is acceptable in a different range, i.e.

$$b_1 = \begin{cases} \frac{A-B}{2\alpha} + \frac{\hbar}{4}, & \text{if } (A-B) > 0 \\ -\frac{A-B}{2\alpha} + \frac{3\hbar}{4}, & \text{if} (A-B) < 0 \end{cases}. \tag{4.57}$$

Residue at $y = \infty$:
We shall now apply Leacock-Padgett boundary condition to select the correct residue at infinity. Writing y in polar variables as $\rho \exp(i\phi)$, we note that for large y, we have

$$(\rho_1\rho_2)^{1/2} e^{i(\phi_1+\phi_2)/2} \rightarrow \rho e^{i\phi} = y. \tag{4.58}$$

because $\rho_1, \rho_2 \rightarrow \rho$ and $\phi_1, \phi_2 \rightarrow \phi$. Equation (4.52) then determines the residue of χ_c at infinity to be $\sqrt{E + A^2}/\alpha$ which is positive and we choose the positive sign in (4.44) for the residue of χ_c at infinity.

Thus, from the above discussion, we can see that by applying the boundary condition we can obtain both, the ranges of the parameters and corresponding physically acceptable solutions. We now proceed to obtain the eigenvalues and eigenfunctions for both the phases.

We will now equate the sum of all residues to zero to obtain the energy eigenvalues and eigenfunctions.
Case 1: $(A-B) > 0$, and $(A+B) > 0$:
The energy eigenvalues follow from the requirement that for a meromorphic function, with a finite number of poles, the sum of all residues including the point at infinity must add to zero. This gives the energy eigenvalues

$$E_n = (A+n\alpha)^2 - A^2 . \tag{4.59}$$

Energy Eigenfunctions
We now turn to the eigenfunctions. Using χ from (4.35), with the above values of the residues in (4.32), the equation for $P_n(y)$ assumes the form

$$(1-y^2)P'' + (\mu - \nu - (\mu + \nu + 1)y)P' + n(n + \mu + \nu)P_n(y) = 0 \tag{4.60}$$

with

$$\mu = \frac{A-B}{\alpha}, \quad \nu = \frac{A+B}{\alpha}, \tag{4.61}$$

and is seen to be the Jacobi differential equation and hence $P_n(y)$ coincides with $P_n^{\mu-\frac{1}{2}, \nu-\frac{1}{2}}(y)$ and the bound state wave function, for the case in which SUSY is exact, is

$$\psi_n(y) = v(1-y)^{\mu/2}(1+y)^{v/2} P_n^{\mu-\frac{1}{2},v-\frac{1}{2}}(y).$$
(4.62)

Case 2 $(A - B) > 0$, and $(A + B) < 0$):
The right choice of the residues for this range, as shown above is,

$$b_1 = \frac{(A-B)}{2\alpha} + \frac{\hbar}{4}, \qquad b_1' = -\frac{A+B}{2\alpha} + \frac{3\hbar}{4}.$$
(4.63)

Substituting χ from (4.35), with the above values of the residues in (4.32) with $\hbar = 1$, one obtains the energy eigenvalue as

$$E_n = \left(B - \left(n + \frac{1}{2}\right)\alpha\right)^2 - A^2$$
(4.64)

Energy Eigenfunctions Substituting the eigenvalues, the differential equation for $P_n(y)$ turns out to be

$$(1-y^2)P_n''(y) + ((v - \mu + 1) - y(v + \mu + 2))P_n'(y)$$
$$+ n(n + \mu + v + 1)P_n(y) = 0$$
(4.65)

and the bound state wave function is found to be

$$\psi_n(y) = v(1-y)^{\mu/2}(1+y)^{\frac{v}{2}+\frac{1}{2}} P_n^{v+\frac{1}{2},\mu-\frac{1}{2}}(y)$$
(4.66)

where

$$\mu = \frac{A-B}{\alpha}, \qquad v = \frac{1}{2} - \frac{A+B}{\alpha}.$$
(4.67)

Thus, we see that the QHJ formalism in one dimension gives the accurate expressions for the bound state wave functions, when there are different phases of SUSY. It may be remarked here that, in the range $A - B < 0$, $A + B < 0$, SUSY is exact but the roles of H_- and H_+ are interchanged. In the range $A - B > 0$, $A + B < 0$, SUSY is again broken.

We close this section by emphasizing that the boundary condition, as proposed by Leacock and Padgett, automatically and correctly picks up the ranges and gives the corresponding solutions.

4.4   Quasi-Exactly Solvable Models

4.4.1   Introduction to QES Models

In this section we study the sextic QES model in one dimension. The QES models are the models for which a part of the bound state energy spectrum and corresponding eigenfunctions can be obtained exactly.

The first instance of a QES model was found by V. Singh et al. who constructed explicit solution of the ground state of the sextic anharmonic oscillator model [6]. This was followed by several other constructions along the same lines. The term "quasi-exact solvability" was introduced by Turbiner and Ushveridze in context of two dimensional models [7]. A large number of investigations of QES models have been reported during eighties and nineties. For a detailed historical account of developments in the area of QES models and references, we refer the reader to the monograph by Ushveridze [5].

In order that a part of the spectrum be obtained exactly, the potential parameters must satisfy a condition known as the condition for quasi-exact solvability. Within the QHJ approach, as applied to ES models, it is not clear how such a condition can arise and why only a part of the spectrum is exactly solvable. In this section we present a resolution of the problems mentioned above in the QHJ formalism. The details of the QHJ treatment of the sextic oscillator, and of other QES models can be found in [2, 8]

In order to study the QES models within the QHJ formalism one needs to have information of singularities of the QMF. This, in general, is not very easy to obtain except for some simple cases like harmonic oscillator and hydrogen atom problems. In the limit $\hbar \to 0$, the QMF $p(x, E)$ formally goes over to $p_{cl}(x, E) = \sqrt{E - V(x)}$. which will, in general, have several branch points. This is an indicates the singularity structure of $p(x, E)$ is complicated. **In order to make progress, we make a simplifying assumption that the point at infinity is an isolated singular point and more specifically it is a pole of some finite order**. We will proceed as in the case of ES models and work out the consequences of the exact quantization condition

$$\frac{1}{(2\pi)} \oint p \mathrm{d}x = n\hbar.$$ (4.68)

For all QES potential models studied by us, it is found that QES condition follows from the LPEQC, Eq. (4.68) and our assumption about the point at infinity.

A further analysis gives us a set of algebraic equations which when solved give us the spectrum and corresponding eigenfunctions.

4.4.2 A List of QES Potentials

A list of potentials studied and the condition of quasi-exact solvability in each case are given below.

The potentials are:

1. Sextic oscillator:

$$V(x) = ax^2 + \beta x^4 + \gamma x^6, \qquad \gamma > 0.$$ (4.69)

The QES condition for the sextic oscillator is

$$\frac{1}{\sqrt{\gamma}}\left(\frac{\beta^2}{4\gamma} - \alpha\right) = 3 + 2n, \qquad n = \text{integer} \tag{4.70}$$

2. Sextic oscillator with centrifugal barrier:

$$V(x) = 4\left(s - \frac{1}{4}\right)\left(s - \frac{3}{4}\right) + \left[b^2 - 4a\left(s + \frac{1}{2} + \mu\right)\right]x^2 + 2abx^4 + a^2x^6.$$
$$\tag{4.71}$$

The QES condition is $\mu = \text{integer}$.
3. Circular potential:

$$V(x) = \frac{A}{\sin^2 x} + \frac{B}{\cos^2 x} + C\sin^2 x - D\sin^4 x, \tag{4.72}$$

where

$$A = 4\left(s_1 - \frac{1}{4}\right)\left(s_1 - \frac{3}{4}\right), \tag{4.73}$$

$$B = 4\left(s_2 - \frac{1}{4}\right)\left(s_2 - \frac{3}{4}\right), \tag{4.74}$$

$$C = q_1^2 + 4q_1(s_1 + s_2 + \mu), \tag{4.75}$$

$$D = q_1^2. \tag{4.76}$$

and the QES condition is $\mu = \text{integer}$.
4. Hyperbolic potential:

$$V(x) = -\frac{A}{\cosh^2 x} + \frac{B}{\sinh^2 x} - C\cosh^2 x + D\cosh^4 x, \tag{4.77}$$

where the parameters A, B, C, D have the same expressions as for the circular oscillator and the QES condition is $\mu = \text{integer}$.
5.

$$V(x) = A\sinh^2\sqrt{\nu}x + B\sinh\sqrt{\nu}\,x + C\tanh\sqrt{\nu}x\,\text{sech}\sqrt{\nu}x + D\text{sech}^2\sqrt{\nu}x \tag{4.78}$$

The QES condition is

$$\left[B \pm 2(n+1)\sqrt{\nu A}\right]^4 + A(4D - \nu)\left[B \pm 2(n+1)\sqrt{\nu A}\right]^2 - 4A^2C^2 = 0 \tag{4.79}$$

6.

$$V(x) = A \cosh^2 \sqrt{v}x + B \cosh \sqrt{v}\,x + C \coth \sqrt{v}x \, \mathrm{csch}\sqrt{v}x + D\mathrm{csch}^2\sqrt{v}x$$
(4.80)

The QES condition is

$$\left[B \pm 2(n+1)\sqrt{vA}\right]^4 - A(4D+v)\left[B \pm 2(n+1)\sqrt{vA}\right]^2 + 4A^2C^2 = 0$$
(4.81)

4.4.3 QHJ Analysis of Sextic Oscillator

We begin with the sextic oscillator potential

$$V(x) = \alpha x^2 + \beta x^4 + \gamma x^6.$$
(4.82)

We will show that the QES condition (4.70) of the previous section follows naturally from the LPEQC. With $q \equiv ip$ the QHJ equation takes the form

$$q^2 + q' - (\alpha x^2 + \beta x^4 + \gamma x^6) + E = 0.$$
(4.83)

The QMF q has no fixed singular points. It has a finite number of moving poles and the point at infinity has been assumed to be a pole of some finite order. It then follows that the QMF is a rational function and hence LPEQC is equivalent to the sum of all residues at all poles, including the point at infinity, is zero. Therefore we get

$$v + \rho_\infty = 0,$$
(4.84)

where v is the total number of moving poles and ρ_∞ is the residue of q at infinity. It will be seen in the following that, for the QES models the moving poles exist on the real axis, as well off the real axis. This is in contrast with the ES models, where the QMF had moving poles on the real line alone and their location coincided with the nodes of the wave function.

4.4.4 The Residue at Infinity

To compute the residue, ρ_∞, at infinity, we change the variable to $\zeta = 1/x$ and with $\tilde{q}(\zeta, E) = q(x, E)\big|_{x \to 1/\zeta}$, the QHJ equation takes the form

$$\tilde{q}(\zeta, E)^2 - \zeta^2 \tilde{q}(\zeta, E) = \left[\frac{\alpha}{\zeta^2} + \frac{\beta}{\zeta^4} + \frac{\gamma}{\zeta^6}\right] - E, \qquad \gamma > 0. \qquad (4.85)$$

As is easily guessed, the negative powers of ζ in Laurent expansion of $\tilde{q}(\zeta, E)$ will terminate at ζ^{-3}. Therefore, we write

$$\tilde{q}(\zeta, E) = \frac{b_3}{\zeta^3} + \frac{b_2}{\zeta^2} + \frac{b_1}{\zeta} + a_0 + a_1\zeta + \cdots . \qquad (4.86)$$

Since the potential $V(x)$ is an even function of x, the bound state wave functions are nondegenerate and have definite parity, i.e. $\psi(-x) = \pm\psi(x)$ and $q(x) = (1/\psi(x))(d\psi/dx)$ will be an odd function of x. Therefore, it follows that $b_2 = 0$ and $a_n = 0$ for even n. Thus (4.86) takes the form

$$\tilde{q}(\zeta, E) = \frac{b_3}{\zeta^3} + \frac{b_1}{\zeta} + a_1\zeta + \cdots \qquad (4.87)$$

The residue at infinity, ρ_∞, being equal to negative of coefficient of ζ, is seen to be $\rho_\infty = -a_1$. The quantization condition (4.84) gives $a_1 = \nu$.

Substituting (4.87) in the QHJ (4.85), comparing coefficients of powers $\frac{1}{\zeta^6}, \frac{1}{\zeta^4}$ and $\frac{1}{\zeta^2}$ gives

$$b_3^2 = \gamma , \qquad (4.88)$$
$$2b_1 b_3 = \beta, \qquad (4.89)$$
$$b_1^2 + 2b_3 a_1 + 3b_3 = \alpha. \qquad (4.90)$$

Of the two values of $b_3 = \pm i\sqrt{\gamma}$, the choice $b_3 = -\sqrt{\gamma}$ leads to a square integrable wave function. With this choice (4.89) gives

$$b_1 = -\beta/2\sqrt{\gamma} . \qquad (4.91)$$

Substituting the values $b_3 = -\sqrt{\gamma}$ and $a_1 = \nu$ in (4.90) leads to

$$\frac{\beta^2}{4\gamma} - (2\nu + 3)\sqrt{\gamma} = \alpha. \qquad (4.92)$$

This equation is just the QES condition. To compare with the result of Ushveridze et al., we write $\gamma = a^2$, $\beta = 2ab$ in (4.92) and obtain

$$\alpha = b^2 - (2\nu + 3)a. \qquad (4.93)$$

The sextic oscillator potential takes the form

$$V(x) = (b^2 - (2\nu + 3)a)x^2 + 2abx^4 + a^2x^6 \qquad (4.94)$$

in agreement with the corresponding equations in [5].

To summarize, we have the following remarkable results for the QES models.

- We started with potential expression (4.82) and the QHJ analysis led to the QES condition, Eq. (4.92), between parameters α, β, γ and each ν corresponds to a different potential. Thus we are now concerned with a family of potentials $\{V_\nu(x)|\nu = 0, 1, \ldots\}$.
- The point at infinity is an isolated singular point.
- *The LPEQC does not give energy eigenvalues; it explains the QES condition.*
- The number ν in the LPEQC equals the total number of the moving poles of QMF, same as the total number of *complex* zeros of the eigenfunctions.
- In contrast with the ES models, as will see below, the energy eigenfunctions have complex zeros.

The above results are totally unexpected and hold for all the other QES models too. For further details we refer to [2, 8].

4.4.5 Form of the QMF and Wave Function

The large x behaviour of $q(x)$ has been determined in the previous subsection. In terms of parameters a, b we have

$$q(x, E) = b_3 x^3 + b_1 x + O(1/x) = -ax^3 - bx + O(1/x). \tag{4.95}$$

The form of the QMF can now be written down using Liouville theorem

$$q(x) = \sum_{k=0}^{n} \frac{1}{x - x_k} - (ax^3 + bx) \tag{4.96}$$

$$= \frac{P_\nu'(x)}{P_\nu(x)} - (ax^3 + bx). \tag{4.97}$$

where $P_\nu(x)$ is a polynomial of degree ν and the term $bx + ax^3$ corresponds to the behaviour of $q(x)$ at infinity. The eigenfunctions, apart from an overall normalization constant, have the form

$$\psi(x) = P_\nu(x) \exp(-ax^4/4 - bx^2/2). \tag{4.98}$$

The above expression is only the form of the eigenfunction $\psi(x)$, the polynomial $P_\nu(x)$ is yet to be determined. The polynomial $P_\nu(x)$ is determined by substituting the above expression in the QHJ equation, which gives

$$P_\nu'' - P_\nu'(2ax^3 + 2bx) - P_\nu(b - E - 2a\nu x^2) = 0. \tag{4.99}$$

Writing $P_\nu(x) = c_0 + c_1 x + \cdots + c_\nu x^\nu$, we are led to a set of linear homogeneous equations in the coefficients of the polynomial $P_\nu(x)$ and (4.99) takes the form of a matrix equation

$$\underline{\mathcal{D}} \begin{pmatrix} c_0 \\ \cdots \\ c_\nu \end{pmatrix} = 0, \tag{4.100}$$

where $\underline{\mathcal{D}}$ is an $(\nu + 1) \times (\nu + 1)$ matrix. For a nontrivial solution, the det $\underline{D}$ must be 0. The sextic oscillator potential is an even function of x. Hence the polynomial $P_\nu(x)$ must be either an odd or an even polynomial. Hence a subset, the even or the odd coefficients, vanish. The number N of remaining undetermined coefficients is seen to be

$$N = \begin{cases} \nu/2 + 1, & \text{for even } \nu \\ (\nu + 1)/2 + 1, & \text{for odd } N. \end{cases} \tag{4.101}$$

Equation (4.100) becomes linear homogeneous equation in N unknown constants. The requirement that a nontrivial solution exists leads to N energy eigenvalues E_k and corresponding polynomials $P_{\nu,k}$. This leads to the polynomials $P_{\nu,k}(x)$ up to an overall constant. This in turn gives the eigenfunctions as

$$\psi_k(x) = P_{\nu,k}(x) \exp(-ax^4/4 - bx^2/2), k = 1, \ldots \nu. \tag{4.102}$$

The polynomial factor appearing in each of the eigenfunctions $\psi_k, k = 1, \ldots, k$ has degree ν. We will now give explicit solutions for $\nu = 0, 1, 2$

Cases $\nu = 0, 1$
The sextic oscillator potential depends on ν via the QES condition. Thus we have a family of potentials $V_\nu(x)$, $\nu = 0, 1, 2, \ldots$ given by Eq. (4.94).
   For $\nu = 0$, the polynomial $P_\nu(x)$ reduces to a constant c_0 and we have

$$p(x, E) = iax^3 + ibx \tag{4.103}$$
$$\psi(x) = \exp(-ax^4/4 - bx^2/2). \tag{4.104}$$

Equation (4.99) with $P_\nu(x)$ replaced by a constant gives $E = b$. For $\nu = 1$, we have the polynomial $P_\nu(x) = c_1 x$ and a straightforward analysis leads to the energy eigenvalue and eigenfunction

$$E = 3b, \qquad \psi(x) = Nx \exp(-ax^4/4 - bx^2/2) \tag{4.105}$$

We will derive explicit form of the eigenfunctions for $\nu = 0, 1, 2$. Later we will discuss the general form of the eigenfunction for arbitrary ν. The general strategy for obtaining the eigenfunctions is the same as discussed for ES models in Ch.3.

Case $v = 2$:

We seek a solution of the form (4.98) with $P_v(x)$ as a second degree polynomial. Substituting $P(x)$ as

$$P_v(x) = c_0 + c_1 x + c_2 x^2 . \tag{4.106}$$

Using above equation in (4.98) and comparing different powers of x gives

$$c_1 = 0, \tag{4.107}$$
$$4ac_0 - c_2(5b - E) = 0, \tag{4.108}$$
$$c_0(b - E) - 2c_2 = 0. \tag{4.109}$$

The last two equations have nontrivial solution for c_0 and c_2 only if

$$\begin{vmatrix} 4a & (5b - E) \\ (b - E) & -2 \end{vmatrix} = 0 . \tag{4.110}$$

This gives two energy eigenvalues

$$E = 3b \pm 2\sqrt{b^2 + 2a} . \tag{4.111}$$

To get the eigenfunction, we compute c_1 and c_2 from equation (4.108) and (4.109) and use

$$p(x) = -i \frac{2c_2 x}{c_0 + c_2 x^2} + iax^3 + ibx . \tag{4.112}$$

Therefore, the eigenfunction is given by:

$$\psi(x) = C \exp\left(\int q(x) dx \right) \tag{4.113}$$
$$\psi(x) = C(c_0 + c_2 x^2) \exp\left(-ax^4/4 - bx^2/2 \right) \tag{4.114}$$

where C is the normalizing factor. The value of c_0 is given by

$$c_0 = \frac{5b - E}{4a} . \tag{4.115}$$

Replacing the value of α_0 and energy value E in the above equation one gets the expression for eigenfunction as

$$\psi(x) = C\left(\frac{E - 5b}{4a} \right) \left[b \pm \sqrt{b^2 + 2a} x^2 - 1 \right] \exp\left(-ax^4/4 - bx^2/2 \right) . \tag{4.116}$$

The eigenfunctions and eigenvalues explicitly obtained for the cases $n = 0, 1$ and 2 agree with the known results [5]

Table 4.1 A comparison of zeros of eigenfunctions of QES and ES models

	QES models	ES models
1	Analytic solutions only if the potential parameters obey QES condition	No condition is required
2	Analytic solutions known only for some states	Solutions are known for all states
3	The wave function has complex zeros	The wave function has no complex zeros
4	The number of real zeros increases and the number of complex zeros decreases with energy and the total number of zeros remains unchanged	Only real zeros whose number keeps increasing with energy
5	As energy increases, complex zeros move to the real axis	No such movement is present

4.4.6 Properties of the Solutions

Recall that all N eigenfunctions $\psi_{v,k}(x)$, $k = 1, \ldots, N$, corresponding to a value v, have a polynomial factor with the same degree v for all k. Hence all the known eigenfunctions have the same number of zeros given by v. Among these zeros, some will be on the real axis and others will be on complex locations off the real axis. If the energy levels are arranged according to increasing energy, the number of zeros on the real axis (nodes) will increase and the number of complex zeros will decrease. This feature has been found to be a general property of the QES models and can be explicitly checked by writing the wave functions for $v > 2$.

A comparison of properties of solutions of QES models with those of ES models is given in table below and schematically represented in Fig. 4.4.

The moving poles of QMF for QES models show a very different pattern for ES and QES models. The location of moving poles coincides with the nodes of the energy eigenfunctions. The figure below sketches a comparison of wave function nodes of typical ES and QES models for first six levels.

As far as the states, for which the analytic solutions cannot be found are concerned, the oscillation theorem gives the information about the real zeros only and no general statement can be made about the complex zeros.

4.5 Band Edges for a Periodic Potential

In this and the next section we take up investigation of a class of periodic potentials related to the Jacobi elliptic functions $\text{sn}(x, m)$, $\text{cn}(x, m)$, $\text{dn}(x, m)$. We first present the QHJ analysis of Lamé potential which is exactly solvable and in the next section we will take up the study a QES periodic model. For general properties and a standard treatment of periodic potentials, see Arscot [9]. For our presentation we shall closely follow [1, 10, 11]. The Lamé potential will be written in the form

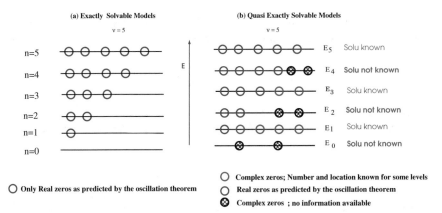

Fig. 4.4 Zeros of eigenfunctions of ES and of QES models

$$V(x) = j(j+1)m \operatorname{sn}^2(x, m). \tag{4.117}$$

With a change of variable to $t = \operatorname{sn}(x, m)$, the QHJ equation

$$q^2 + q' + E - V(x) = 0 \tag{4.118}$$

takes the form

$$\chi^2 + \frac{d\chi}{dt} + \frac{(mt)^2 + 2m}{4(1 - mt^2)^2} + \frac{t^2 + 2}{4(1 - t^2)^2} + \frac{2E - j(j+1)m^2 - mt^2}{2(1 - t^2)(1 - mt^2)} = 0, \tag{4.119}$$

where

$$q = \sqrt{(1 - t^2)(1 - mt^2)}\,\phi, \quad \phi = \chi + \frac{1}{2}\left(\frac{mt}{1 - mt^2} + \frac{t}{1 - t^2}\right). \tag{4.120}$$

A useful reference for the elliptic functions is [12] where the properties of elliptic functions needed for arriving at (4.119) can be found. Equation (4.119) shows that χ has fixed poles at $t = \pm 1$, and $t = \pm\frac{1}{\sqrt{m}}$. Using ν to denote the number of moving poles, we can write the form of $\chi(t)$ as

$$\chi = \frac{b_1}{t - 1} + \frac{b_2}{t + 1} + \frac{d_1}{t + \frac{1}{\sqrt{m}}} + \frac{d_2}{L + \frac{1}{\sqrt{m}}} + \sum_{k=1}^{\nu} \frac{1}{t - t_k} + Q(t), \tag{4.121}$$

where $Q(t)$ is analytic part of $\chi(t)$. The first four terms represents the fixed poles and the fifth term is a sum of moving poles. From QHJ Eq. (4.119), it is easily seen that $Q(t)$ is bounded at infinity. Hence by Liouville's theorem $Q(t)$ is a constant, say C. Since χ is bounded at infinity, we write

$$\chi = \chi_0 + \frac{\lambda_1}{t} + \frac{\lambda_2}{t^2} + \cdots \qquad (4.122)$$

We note that λ_1 is negative of the residue of $\chi(y)$ at $y = \infty$.

Substituting χ from (4.122) in (4.119) and comparing powers of t on both side we get

$$\lambda_0 = 0, \quad \lambda_1 = j + 1, -j. \qquad (4.123)$$

Since $\chi(t)$ is a rational function, the sum of residues at all poles, including the singularity at infinity, must vanish. This gives

$$2b_1 + 2d_1 + \nu - \lambda_1 = 0. \qquad (4.124)$$

Next we write the sum over all moving poles in (4.121) as

$$\sum_{k=1}^{\nu} \frac{1}{t - t_k} = \frac{P_\nu'(t)}{P_\nu(t)} \qquad (4.125)$$

where $P_\nu(t) = \prod_k (t - t_k)$.

An explicit form of the wave function is obtained using

$$\psi(t) = \exp\left[\int \left(\chi(t) + \frac{1}{2}\frac{mt}{1 - mt^2} + \frac{t}{1 - t^2}\right) dt\right] P_\nu(t)$$

$$= \exp\left[\int \left(\frac{(1 - 4b_1)t}{2(1 - t^2)} + \frac{(1 - 4d_1)mt}{2(1 - mt^2)}\right) dt\right] P_\nu(t). \qquad (4.126)$$

Integrating over t and expressing the answer in terms of the original variable x, we get

$$\psi(x) = (\mathrm{cn}x)^\alpha (\mathrm{dn}x)^\beta P_\nu(\mathrm{sn}x), \qquad (4.127)$$

where

$$\alpha = \frac{4b_1 - 1}{2} \ , \quad \beta = \frac{4d_1 - 1}{2}. \qquad (4.128)$$

In contrast to the case of bound state problems, the wave function is acceptable for all choices of the residues b_1, b_2, d_1 and d_2.

The four cases I, II, III and IV lead to the four sets of forms of the wave functions as listed below.

Case I: $b_1 = \frac{1}{4}, d_1 = \frac{1}{4}, v = j$ $\psi(x) = P_j(\text{sn}(x))$
Case II: $b_1 = \frac{3}{4}, d_1 = \frac{1}{4}, v = j - 1$ $\psi(x) = \text{cn}(x) P_{j-1}(\text{sn}(x))$
Case III: $b_1 = \frac{1}{4}, d_1 = \frac{3}{4}, v = j - 2$ $\psi(x) = \text{dn}(x) P_{j-1}(\text{sn}(x))$
Case IV: $b_1 = \frac{3}{4}, d_1 = \frac{3}{4}$ $\psi(x) = (\text{cn}(x))(\text{dn}(x)) P_{j-2}(\text{sn}(x))$

Since the potential is an even function of x, the polynomial $P_v(x)$ for odd v can have only odd powers and the coefficients of even powers must vanish. Likewise if v is an even integer, the coefficients of odd powers of t must be zero. This restricts the number of unknown parameters, to be denoted by N, in the in the polynomial $P_v(t)$ as follows.

$$N = \begin{cases} (v+1)/2, & \text{if } v \text{ is odd} \\ v/2 + 1, & \text{if } v \text{ is even} \end{cases} \tag{4.129}$$

Substituting $\chi(t)$ from Eq. (4.121) in Eq. (4.119), we get the following equation for the polynomial $P_v(t)$

$$P_v''(t) + 4t \left(\frac{b_1}{t^2 - 1} + \frac{md_1}{mt^2 - 1} \right) P_v' + G(t) P_v(t) = 0 \tag{4.130}$$

where

$$G(t) = \frac{t^2(4b_1^2 - 2b_1 + 1/4) + 1/2 - 2b_1}{(t^2 - 1)^2}$$
$$+ \frac{(mt)^2(4d_1^2 - 2d_1 + 1/4) + m/2 - 2md_1}{(mt^2 - 1)^2}$$
$$+ \frac{2E + 4(m + 1 - \delta) + (16b_1 d_1 - 13)mt^2}{2(t^2 - 1)(mt^2 - 1)} \tag{4.131}$$

The differential equation Eq. (4.130) implies N linear homogeneous equations for the N nonzero coefficients appearing in the polynomial $P_v(t)$ which can be cast in a matrix form as

$$\mathcal{D} \begin{pmatrix} c_{j_1} \\ c_{j_2} \\ \cdots \\ c_{j_N} \end{pmatrix} = 0, \tag{4.132}$$

where $c_{j_1}, c_{j_2}, \ldots, c_{j_N}$ are the nonzero coefficients of, odd (or even) powers of t. Equation (4.132) has non trivial solutions only of the determinant of $N \times N$ matrix $\mathcal{D}$ vanishes. This gives N eigen values of energy and corresponding polynomials $E_k, P_{v,k}(t), k = 1, \ldots, N$. This process gives the solutions for energies and polynomial $P_v(t)$ for each of the four cases listed above.

Table 4.2 Number of solutions N for a given j

Residue sets			j = even, 2η		j = odd, $2\eta + 1$	
Set	b_1	d_1	ν	N	ν	N
I	$\dfrac{1}{4}$	$\dfrac{1}{4}$	2η	$\eta + 1$	$2\eta + 1$	$\eta + 1$
II	$\dfrac{3}{4}$	$\dfrac{1}{4}$	$2\eta - 1$	η	2η	$\eta + 1$
III	$\dfrac{1}{4}$	$\dfrac{3}{4}$	$2\eta - 1$	η	2η	$\eta + 1$
IV	$\dfrac{3}{4}$	$\dfrac{3}{4}$	$2\eta - 2$	η	$2\eta - 1$	η

Counting the number of unknown parameters N and its values for odd and even j are listed in separate columns of Table 4.2 .

Equation (4.124) restricts the combinations of residues that are permitted. Noting that $\lambda - 2b_1 - 2d_1(= \nu)$ should be positive, we must choose $\lambda_1 = j + 1$, discarding the value $\lambda_1 = -j$ we get the four allowed combinations of b_1, d_1 and corresponding values of ν are given in the Table 4.2

The four cases in the table give four solutions which are in agreement with known results for band edge wave functions [9].

4.5.1 Explicit Solution for $j = 2$

We will now give explicit solution of the Lame potential for $j = 2$. We need to find, for $\nu = 2$, the polynomial part, $P_\nu(x)$, of the solution of $\psi(x)$ given by Eq. (4.127). This needs to be done for the four combination of residues listed in Table 4.2.

Set I
Taking the values of residues in the first row of the table, for $j = 2$, the degree of the polynomial $P_\nu(t)$ is $\nu = 2$. Therefore we take $P_2(t) = A + Bt + Ct^2$ and substitute it in Eq. (4.130). We equate different powers of t to zero. Coefficient of t^4 vanishes. Equating the coefficient of t^3 term to zero implies $B = 0$. The remaining two coefficients A, C satisfy the following equation

$$\begin{pmatrix} -6m & (E - 2\delta) - (2m + 2) \\ (E - 2\delta) + (2m + 2) & 2 \end{pmatrix} \begin{pmatrix} A \\ C \end{pmatrix} = 0. \qquad (4.133)$$

For nontrivial solutions, the determinant of the matrix in the above equation must be zero. Using $\delta = \sqrt{1 - m + m^2}$, we get the energy eigenvalues

$$E_1^I = 4\delta \qquad E_2^I = 0. \qquad (4.134)$$

and the corresponding polynomials are found to be

$$P^I_{2,1}(t) = m + 1 - \delta - 3mt^2, \qquad P^I_{2,2}(t) = m + 1 + \delta - 3mt^2. \qquad (4.135)$$

The corresponding solutions are obtained from (4.127) by setting $t = \mathrm{sn}(x)$ and $\alpha = \beta = 0$:

$$\psi^I_1(x) = m + 1 - \delta - 3m\,\mathrm{sn}(x)^2, \qquad \psi^I_2(x) = m + 1 + \delta - 3m\,\mathrm{sn}(x)^2. \quad (4.136)$$

For the other cases listed in the rows 2,3,4, the polynomial $P_\nu(t)$ is an odd polynomial of degree 1. Writing $P(\nu(t)) = Ct$ Eq. (4.130) gives nonzero value for C only for energy equal an eigenvalue. The final answers energy value and wave functions for the band edges, for of residue sets II–IV, are as follows.

Set II $E^{II} = 2\delta - m + 2$, $\quad\psi^{II}(x) = \mathrm{cn}x\,\mathrm{sn}x$.
Set III $E^{III} = 2\delta + 2m - 1$, $\quad\psi^{III}(x) = \mathrm{dn}x\,\mathrm{sn}x$.
Set IV $E^{IV} = 2\delta - m - 1$, $\quad\psi^{IV}(x) = \mathrm{cn}x\,\mathrm{dn}x$.

For Lamé and associated Lamè potentials we find that there are four groups of solutions corresponding to the four possible combinations of residues.

For a given combination of the residues, there are, in general, several independent solutions. These independent solutions, for each combination of residues, contain a polynomial factor of the same degree and, therefore, have the same number of total real and complex zeros. This pattern is similar to what was found for the QES sextic oscillator. For further discussion of explicit solutions see [1, 11]. Using the notation $p = a(a + 1), q = b(b + 1)$, the potential,

$$V(x) = pm\,\mathrm{sn}^2(x, m) + qm\,\mathrm{cn}^2(x, m)/\mathrm{dn}^2(x, m)$$

is exactly solvable for $a = b = j$ and QES for $a \neq b$.

More details of QHJ study of Lamé and associated Lamé can be found in [1, 11, 13].

4.6   A QES Periodic Potential

In this section we shall obtain the energy eigenfunctions and eigenvalues for the associated Lamé potential

$$V(x) = a(a + 1)m\,\mathrm{sm}^2(x, m) + b(b + 1)m\left(\frac{\mathrm{cn}(x, m)}{\mathrm{dn}(x, m)}\right)^2, \quad a \neq b, \qquad (4.137)$$

which is a periodic QES potential . A QHJ treatment of this model will be presented closely following [13].

Performing a change of variable from x to $y = sn(x)$ we arrive at the QHJ equation

$$\chi^2 + \frac{d\chi}{dy} + \frac{m^2 y^2 + 2m(1 - 2b(b+1))}{4(1 - mt^2)^2} + \frac{2 + y^2}{4(1 - y^2)^2} + \frac{2E - my^2(1 - 2a(a+1))}{2(1 - y^2)(1 - my^2))} = 0.$$
(4.138)

where

$$\chi = \phi - \frac{1}{2}\left(\frac{mt}{1 - mt^2} + \frac{t}{1 - t^2}\right), \quad \phi = \frac{q}{\sqrt{(1 - t^2)(1 - mt^2)}},$$
(4.139)

From QHJ Eq. (4.138) for $\chi(y)$, it is seen that χ has fixed poles at $y = \pm 1$, $y = \pm 1/\sqrt{m}$. Assuming a finite number of moving poles, we arrive at the following form for $\chi(y)$.

$$\chi(y) = \frac{b_1}{y - 1} + \frac{b_1'}{t + 1} + \frac{d_1}{y - 1/\sqrt{m}} + \frac{d_1'}{t + 1/\sqrt{m}} + \sum_n \frac{1}{P_n(y)} \frac{dP_n(y)}{dy} + Q(y).$$
(4.140)

Here $P_\nu(y)$ is a polynomial of degree ν and $Q(y)$ is an analytic function of y which is to be fixed by making use of behaviour of $\chi(y)$ for large y. Appealing to Liouville's theorem, Eq. (4.138) suggests $Q(y) = C$ as $\chi(y) \to 0$ for large y. The residues b_1, b_1', d_1, d_1' at poles $\pm 1, \pm 1/\sqrt{m}$ are fixed as before by substituting Laurent expansions about the corresponding points in the QHJ equation (4.138). This gives the residues as

$$b_1 = \frac{3}{4}, \frac{1}{4}, \quad b_1' = \frac{3}{4}, \frac{1}{4}$$
(4.141)

$$d_1 = \frac{3}{4} + \frac{b}{2}, \frac{d_1}{4} - \frac{b}{2}, \quad d_1' = \frac{3}{4} + \frac{b}{2}\frac{1}{4} - \frac{b}{2}.$$
(4.142)

The residue at infinity is easily found to be given by one of the two values

$$\alpha = 1 + a, -a.$$
(4.143)

Symmetry of the QHJ equation (4.138) under $y \to -y$ implies that we must take

$$b_1 = b_1' \text{ and } d_1 = d_1'.$$
(4.144)

The requirement that the sum of all residues must vanish for a meromorphic function with a finite number of poles becomes

$$2b_1 + 2d_1 + n = \alpha.$$
(4.145)

This equation restricts the allowed combinations of b_1, d_1 and α. The associated Lamé problem has a symmetry of the problem under $a \leftrightarrow -(a+1)$ and

Table 4.3 Residues and QES condition

	b_1	d_1	QES condition
Case 1	$\dfrac{3}{4}$	$\dfrac{3}{4} + \dfrac{b}{2}$	$a - b = n + 2$
Case 2	$\dfrac{3}{4}$	$\dfrac{1}{4} - \dfrac{b}{2}$	$a + b = n + 1$
Case 3	$\dfrac{1}{4}$	$\dfrac{3}{4} + \dfrac{b}{2}$	$a - b = n + 1$
Case 4	$\dfrac{1}{4}$	$\dfrac{1}{4} - \dfrac{b}{2}$	$a + b = n$

$b \leftrightarrow -(b+1)$. Therefore, we can assume $a > 0, b > 0$ without loss of generality. This gives us the following four cases of allowed combinations of residues (Table 4.3).

The band edge energy eigenvalues and eigenfunctions can now be easily determined using the QHJ equation (4.138). The details can be found in [1] and [13].

4.7   Some Observations

We close this study of different models studied in this chapter by highlighting differences with ES models in the previous chapter.

1. The Scarf potential has two different spectra for different ranges of coupling constants. The spectra and corresponding ranges of the coupling constant come out automatically when LPBC is carefully implemented. Similar comments apply to the Scarf-I potential with two phases of SUSY.
2. In case of QES models, LPEQC does not lead to energy levels but to the QES condition . The energy levels and eigenfunctions are obtained correctly by making further detailed analysis of QHJ equation.
3. The wave functions of all the QES models studied have complex zeros in addition to the real nodes. This is in contrast with eigenfunctions of ES having only real zeros as given by oscillation theorems.

References

1. Sree Ranjani, S.: Quantum Hamilton Jacobi solution for spectra of one dimensional potentials with special properties, Ph. D. Thesis submitted to University of Hyderabad. arXiv:quant-ph/0410.8036v1 (2004)
2. Geojo, K.G.: Quantum Hamilton Jacobi study of wave functions and energy spectrum of solvable and quasi-exactly solvable models, Ph. D. Thesis submitted to University of Hyderabad. arXiv:quant-ph/0410008v1 (2004)

3. Cooper, F., et al.: Supersymmetry and quantum mechanics. Phys. Rep. **251**, 267–385 (1995)
4. Scarf, F.L.: New soluble energy band problem. Phys. Rev. **112**, 1137–1140 (1958)
5. Ushveridze, Alexander: Quasi-Exactly Solvable Models in Quantum Mechanics. Institute of Physics Publishing, Bristol and Philadelphia (1993)
6. Singh, V., et al.: Anharmonic oscillator and the analytic theory of continued fractions. Phys. Rev. D **18**, 1901–1908 (1978)
7. Turbiner, A.V., Ushveridze, A.G.: Preprint ITEP-169 Moscow (1986)
8. Geojo, K.G., et al.: A study of exactly solvable models within the quantum Hamilton-Jacobi formalism. J. Phys. Math. Gen. A **36**, 4591–4598 (2003)
9. Arscot, F.M.: Periodic Differential Equations. Elsevier Science, Amsterdam (2014)
10. Sree Ranjani, S. et al.: Bound state wave functions through QHJ formalism. Mod. Phys. Lett. A **19**, 1457–1468 (2004)
11. Sree Ranjani, S. et al.: Band edge eigenfunctions and eigenvalues for periodic potentials through the quantum Hamilton-Jacobi formalism. Int. J. Mod. Phys. **19A**, 2047–2058 (2004)
12. Harris, H.: Lectures on Theory of Elliptic Functions. Dover Publications, New York (2004)
13. Sree Ranjani, S. et al.: Periodic quasi exactly solvable models. Int. J. Theor. Phys. **44**, 1167-1176 (2005). arxiv; quant-ph/0403196 VI, 2nd March 2004

Chapter 5
Rational Extensions

5.1 Introduction

One-dimensional ES models, some of which the reader has encountered in the previous chapters, are well studied and appear as standard examples in text books. While they are effective in conveying the physical and mathematical aspects of quantum mechanics to novices, they also provide useful testing ground for trying out new analytical and numerical methods developed to solve potentials which are not ES. The number of ES models has been more or less fixed for many years till the construction of the rational potentials involving EOPs in their solutions. In 2009, Gomez-Ullate *et.al.,* in their fascinating paper [1] introduced these EOPs, which led to the construction of the new rational potentials by Quesne [2], which are ES. Interestingly the new potentials and their solutions are intimately connected to the already existing ES models and their solutions which involve COPs [3–5]. In [6] we have analyzed these new potentials and obtained their solutions. From here on we refer to these older potentials as regular potentials, while the new potentials will be referred to as the rationally extended potentials or simply as the extended potentials.

Over the years, many methods have been employed to construct these potentials [7–16]. For more details about the properties and the various methods employed in constructing these potentials, we refer to [17, 18] and references therein. In this chapter, we will employ the QHJ formalism to construct these extended potentials demanding them to be shape invariant [19]. In comparison to the existing methods, the formalism offers a very transparent and simple method to construct these potentials. In fact the existence of these potentials appears very natural, when looked from the scheme of QHJ formalism.

In addition, we also discuss the property of shape invariance (SI) [20], satisfied by most ES models in one dimension, which is associated with supersymmetry in Quantum Mechanics [21–23]. We define SI to be a property of the potential and present the necessary conditions required [24].

© The Author(s), under exclusive license to Springer Nature Switzerland AG 2022
A. K. Kapoor et al., *Quantum Hamilton-Jacobi Formalism,*
SpringerBriefs in Physics, https://doi.org/10.1007/978-3-031-10624-8_5

5.2   About ES Potentials and Orthogonal Polynomial Connection

Most of the regular ES potentials, one dimensional models as well as separable models in higher dimensions, are listed in [3–5]. It is a common knowledge that

1. the Schrödinger equation for these ES potentials reduces to the Sturm-Liouville [25] type equation after a suitable change of variable.
2. The eigenfunctions of these potentials are in terms of the COPs [26].

The latter statement is due to Bochner, whose result [27, 28] characterizes the COPs as the polynomial solutions of the Sturm-Liouville problem. The precise statement of the result as stated in [1] is:

If an infinite sequence of polynomials, $\{P_n(x)\}_{n=0}^{\infty}$, satisfies a second order differential equation of the form

$$p(x)P_n''(x) + q(x)P_n'(x) + r(x)P_n(x) = \lambda_n P_n(x), n = 0, 1, 2, \ldots, \qquad (5.1)$$

then $p(x)$, $q(x)$ and $r(x)$ must be polynomials of degree 2, 1 and 0 respectively. In addition, if the sequence $P_n(x)$ is an orthogonal polynomial system, then it has to be (up to an affine transformation of x) one of the COP systems of Jacobi, Laguerre or Hermite [1]. In the QHJ analysis, for all the ES models, the moving pole part of the QMF turns out to be the logarithmic derivative of the COPs.

5.3   Exceptional Orthogonal Polynomials

In [1], Gomez-Ullate et al. successfully argued that the Sturm-Liouville theory does not impose the requirement that the polynomial sequence should start with a constant. They demonstrated that new sequences of orthogonal polynomials can be constructed, which start with a polynomial of degree m with $m = 1, 2, 3 \ldots$. These were named as the EOPs. These appear as the solutions of the Sturm-Liouville type equations, thus extending the Bochner's result to encompass the new EOPs as the solutions of the Sturm-Liouville equation.

Over the last decade, generalized families of exceptional Hermite, Laguerre and Jacobi have been constructed and their properties have been well studied [29–34]. Similar rational polynomial sequences related to other orthogonal polynomials have also been constructed [35–39]. These extended polynomials have similar characteristics as the COPs. Here we will throw some light on some of the characteristics of EOPs like the codimension index, zeros of the polynomials and the weight functions in which they differ from their classical counterparts.

Codimension index: One of the major deviations is that while any sequence of the COPs, $P_n(x)$, starts with $P_0(x) = 1$, any EOP sequence $P_{n,m}(x)$ starts with a mth degree polynomial. The index m is known as the codimension index and takes integer values 1, 2, 3, ... In addition, it gives the number of gaps in the polynomial sequence [40]. Furthermore, for each distinct value of m we get a different EOP sequence and when $m = 0$ all these sequences reduce to the corresponding COP sequence. Note that for both COPs and EOPs the index $n = 0, 1, 2 \ldots$.

Zeros of the orthogonal polynomials: Any n^{th} degree polynomial in the COP sequence has n zeros in the corresponding orthonormality interval. In contrast if we consider a polynomial in an EOP sequence, indexed by the codimension index m, it will have a total of $n + m$ zeroes. Of these, n zeros lie in the orthonormality interval, while m zeros lie outside this interval. The degree of the polynomial is given by the number n. For both the COPs and the EOPs $n = 0, 1, 2, 3, \ldots$

Weight functions: It is well known that the COPs are orthonormal with respect to a weight function. The latter is a well defined, positive normalizable function in the interval of orthonormality. The EOPs are also shown to be orthonormal, but with respect to a rational weight function. The term appearing in the denominator of the weight function has singularities that lie outside the orthonormality interval. This ensures that the weight function is still well behaved and the number of such singularities is equal to the codimension index m.

5.4 Extended Potentials

Just as COPs are related to the solutions of the Schrödinger equation with regular ES potential, the existence of EOPs led to the construction of new rational potentials. These potentials have the form

$$V_{rat}(x) = V_{reg}(x) + \text{Rational terms}, \tag{5.2}$$

where $V_{rat}(x)$ represents the rational potential and $V_{reg}(x)$ represents the regular ES potential. From the expression it is clear as to why the former are also referred to as the rational extensions of $V_{reg}(x)$.
As an example, consider the potential

$$V(r) = \frac{1}{4}\omega^2 r^2 + \frac{\ell(\ell + 1)}{r^2} \tag{5.3}$$

discussed in Chap. 3, Sect. 5.2. It has solutions in terms of the associated Laguerre polynomials. One of the generalized rational potential associated with it is of the form

$$V_{rat}(r) = \frac{1}{4}\omega^2 r^2 + \frac{\ell(\ell+1)}{r^2} + +2\omega r \left[\omega r + \frac{2\ell}{r}\right] \frac{\partial_y L_m^\alpha(y)}{L_m^\alpha(y)}$$
$$-\omega r \partial_y \left[\omega r \frac{\partial_y L_m^\alpha(y)}{L_m^\alpha(y)}\right] + \left(\omega r \frac{\partial_y L_m^\alpha(y)}{L_m^\alpha(y)}\right)^2. \tag{5.4}$$

Here, $y = -\frac{1}{2}\omega r^2$, $\alpha = \ell - 1/2$ and m takes values $1, 2, 3 \ldots$. It has solutions in terms of the $L1$ type exceptional Laguerre polynomials. Note that for $m = 0$, the above potential reduces to (5.3), since $L_0^\alpha(y) = 1$. For $m = 1$ the potential matches with the rational potential constructed in [2]. As is obvious from (5.4), for different m we get distinct rational potentials. This leads to a family of rational extensions indexed by m. In addition to this family of potentials, there are two more families of rational extensions of the radial oscillator.

Our main aim in this chapter is to give details of the construction of the extended potentials using the QHJ formalism. These will be given in Sec 5.9. In order to keep the discussion self contained, a short introduction to supersymmetric quantum mechanics in given in Sec. 5.5, followed by a discussion on SI in Sec. 5.6 and methods to construct new ES potentials, namely the isospectral deformation and isospectral shift deformation in Sec. 5.7. Readers familiar with the concepts can directly go to the SI section and proceed from there.

5.5   Supersymmetric Quantum Mechanics

Supersymmetry (SUSY) was a culmination of efforts in field theory to bring bosons and fermions into a single multiplet [21, 22]. The idea of SUSY was investigated in quantum mechanical models by Witten [23] and others leading to supersymmetric quantum mechanics (SUSYQM). For more details we refer the reader to [4, 5] and references there in.

The novel concepts which SUSYQM introduced into quantum mechanics were the superpotential, the partner potentials and SI [20]. The partner potentials being the artifacts of the bosonic and fermionic degrees of freedom available in the field theory.

In the notation, $\hbar = 2m = 1$, given a function $W(x)$, known as the superpotential, we can construct $V^\pm(x)$ as

$$V^-(x) = W^2(x) - W'(x), \quad V^+(x) = W^2(x) + W'(x). \tag{5.5}$$

The corresponding partner Hamiltonians are defined as

$$H^\pm = -\frac{d^2}{dx^2} + V^\pm(x). \tag{5.6}$$

If one of the two functions $\psi_0^\pm(x) = \exp(\pm \int W(x))$, is square integrable then it is an eigenfunction of the corresponding Hamiltonian H^+ (or H^-). Then $\psi_0^\pm(x)$ are the zero energy wave functions of the Hamiltonians and we say that SUSY is exact. It can then be shown that the partners are isospectral except for the zero energy state. SUSY is said to be broken if both the functions $\psi_0^\pm(x)$ are non-normalizable. In this case, the partners can be shown to be strictly isospectral.

Knowing the superpotential, the Hamiltonians $H^{\pm}(x)$ corresponding to $V^{\pm}(x)$ can be factorized as follows

$$H^{-}(x) = A^{\dagger}A , \quad H^{+}(x) = AA^{\dagger}, \tag{5.7}$$

where A and $A^{\dagger}$, known as the intertwinning operators , are defined as follows

$$A = \frac{d}{dx} + W(x) , \quad A^{\dagger} = -\frac{d}{dx} + W(x). \tag{5.8}$$

The action of the intertwinning operators is contained in the equation below

$$AH^{-} = H^{+}A. \tag{5.9}$$

These operators can be used to obtain the eigenfunctions and the eigenvalues of $H^{\pm}(x)$. For the case of exact SUSY, we have

$$\psi_n^{+}(x) = \frac{1}{\sqrt{E_{n+1}^{-}}} A \psi_{n+1}^{-}(x) , \quad \psi_{n+1}^{-}(x) = \frac{1}{\sqrt{E_n^{+}}} A^{\dagger} \psi_n^{+}(x), \tag{5.10}$$

where $n = 0, 1, 2, \ldots$.

$$E_n^{+} = E_{n+1}^{-} , \quad E_0^{-} = 0. \tag{5.11}$$

It is emphasized that for a given potential, we can construct more than one superpotential. Only one among these will keep the SUSY exact. As seen in Chap. 3, Sect. 5.2, the superpotential is related to the QMF and it corresponds to the solutions of the QHJ equation, which do not have moving poles. Therefore we can construct all the superpotentials associated with $V(x)$ using this relationship.

5.6 Shape Invariance

Given a superpotential $W(x)$, the partner potentials are said to be shape invariant, if

$$V^{+}(x, a_0) = V^{-}(x, a_1) + R, \tag{5.12}$$

where a_0 stands for a set of potential parameters and a_1 is a function of a_0 and R is independent of x [20].

As an example, let us consider the radial oscillator again, whose superpotential is given in Chap. 2 as

$$W(r) = \frac{\omega r}{2} - \frac{l+1}{r}. \tag{5.13}$$

The corresponding partner potentials can be constructed using (5.5) as,

$$V^-(r) = \frac{1}{4}\omega^2 r^2 + \frac{\ell(\ell+1)}{r^2} - \omega\left(\ell + \frac{3}{2}\right) \tag{5.14}$$

$$V^+(r) = \frac{1}{4}\omega^2 r^2 + \frac{(\ell+2)(\ell+1)}{r^2} - \omega\left(\ell + \frac{1}{2}\right), \tag{5.15}$$

which have exact SUSY between them. From the above expressions, it is evident that by shifting the potential parameter $\ell \to \ell + 1$ in $V^-(x)$, we can obtain $V^+(r)$. Clearly the partners satisfy the SI condition (5.12) and in this case $R = -\omega$. The partners are isospectral except for the ground state.

From the above discussion, the SI does not appear to be the property of the potential but depends on the choice of the superpotential. As already seen in Chap. 3, for the radial oscillator potential there are four superpotentials as given in Table 3.1. For a given potential $V(x)$, the superpotential is obtained by solving the following Riccati equation,

$$W^2(x) - W'(x) = E - V(x), \tag{5.16}$$

for some value of E. The superpotential is not uniquely fixed, and we will see in Table 5.1.

Shape Invariance as a Property of the Potential We define SI as a property of the potential without reference to any of the superpotentials associated with it [24].

Definition Consider a potential $V(x, \sigma)$ and the corresponding QHJ equation

$$q^2(x) + \frac{dq(x)}{dx} = V(x, \sigma) - E, \tag{5.17}$$

where σ represents the potential parameters. Let us assume that (5.17) has two solutions $\omega_1(x, \lambda)$ and $\omega_2(x, \mu)$ for some constants E_1 and E_2, which do not have any moving poles. Here λ and μ are functions of the potential parameters.

The potential $V(x)$ will be called shape invariant, if there exists a map

$$\tau : \lambda \to \mu = \tau(\lambda), \tag{5.18}$$

such that

$$\omega_1(x, \lambda) + \omega_2(x, \mu) = 0. \tag{5.19}$$

The above definition implies SI of the potential $V(x, \sigma)$ in the conventional sense.

To see this, we use the solutions $\omega_1(x, \lambda)$ and $\omega_2(x, \mu)$ to construct two potential functions $V_1(x)$ and $V_2(x)$, where

$$V_1(x) = \omega_1^2(x, \lambda) - \omega_1'(x, \lambda) = V(x) - E_1 \quad \text{and} \tag{5.20}$$

$$V_2(x) = \omega_2^2(x, \mu) - \omega_2'(x, \mu) = V(x) - E_2. \tag{5.21}$$

Using the mapping in (5.18), we obtain

$$V_2(x) = \omega_1^2(x, \tau(\lambda)) + \omega_1'(x, \tau(\lambda)) + E_1 - E_2. \tag{5.22}$$

It is obvious that $V_1(x)$ and $V_2(x)$ are shape-invariant partners with $\omega_1(x, \lambda)$ playing the role of the superpotential. Thus the shape invariance of the partners $V_1(x)$ and $V_2(x)$ gives the usual condition

$$V_2(x, \lambda) = V_1(x, \tau(\lambda)) + \text{constant}, \tag{5.23}$$

which can be expressed as

$$w_1^2(x, \tau(\lambda)) + w_1'(x, \tau(\lambda)) = w_1^2(x, \lambda) - w_1'(x, \lambda) + R, \tag{5.24}$$

where $R = E_1 - E_2$. We will use the above form of the SI condition to construct the rational extensions as shown in the following section.

5.7 Construction of New ES Models

In this section we discuss two methods, namely the isospectral deformation and the isospectral shift deformation method used to construct ES potentials. The former is a well-established method in SUSYQM, while the latter is a method we introduced to construct rational extensions using inputs from both SUSYQM and QHJ formalism [19, 24].

5.7.1 Isospectral Deformation

Consider the SI partner potentials, $V^{\pm}(x)$, with exact SUSY between them and the corresponding superpotential $W(x)$. We construct a new superpotential $\tilde{W}(x)$ of the form

$$\tilde{W}(x) = W(x) + \phi(x) \tag{5.25}$$

where the unknown function $\phi(x)$ is determined by demanding that

$$\tilde{V}^+(x) = V^+(x). \tag{5.26}$$

The set of partner potentials corresponding to $\tilde{W}(x)$ are

$$\tilde{V}^-(x) = \tilde{W}^2(x) - \tilde{W}'(x) \ , \quad \tilde{V}^+(x) = \tilde{W}^2(x) + \tilde{W}'(x). \tag{5.27}$$

From (5.26) it is clear that $\tilde{V}^+(x)$ is identical to $V^+(x)$, but $\tilde{V}^-(x)$ will turn out to be an isospectral deformation of $V^-(x)$. Substituting $\tilde{V}^+(x)$ and $V^+(x)$ from (5.5) and (5.27) respectively in (5.26) gives the Riccati equation for the unknown function $\phi(x)$ as

$$\phi^2(x) + 2W(x)\phi(x) + \phi'(x) = 0. \tag{5.28}$$

If $\chi(x) = 1/\phi(x)$, the above equation written in terms of $\chi(x)$, satisfies the Bernoulli equation

$$\chi'(x) = 1 + 2W(x)\chi(x). \tag{5.29}$$

This is a linear equation whose solution can be obtained using an integrating factor and we have

$$\frac{1}{\chi(x)} = \phi(x) = \frac{d}{dx}\ln(I(x) + \lambda), \tag{5.30}$$

where

$$I(x) = \int_{-\infty}^{x} (\psi_0^-)^2(y)dy \tag{5.31}$$

and λ is a constant of integration. Thus, the extended superpotential turns out to be

$$\tilde{W}(x) = W(x) + \frac{d}{dx}\ln(I(x) + \lambda). \tag{5.32}$$

By construction $\tilde{V}^+(x) = V^+(x)$ but $\tilde{V}^-(x) \equiv \tilde{V}^-(x\,;\lambda)$ is given by

$$\tilde{V}(x\,;\lambda) = \tilde{W}^2(x) - \tilde{W}'(x) \tag{5.33}$$

and use of (5.32) gives

$$\tilde{V}^-(x\,;\lambda) = V^-(x) - \frac{4\psi_0^-(x)\psi_0^{-'}(x)}{I(x) + \lambda} + \frac{2(\psi_0^-)^4(x)}{(I(x) + \lambda)^2}. \tag{5.34}$$

The above equation gives us a family of potentials which are indexed by λ. These potentials are strictly isospectral to the potential $V^-(x)$ with their ground states given by [5, 41]

$$\tilde{\psi}_0^-(x) = \frac{\sqrt{\lambda(1 + \lambda)}\psi_0^-(x)}{(I(x) + \lambda)}. \tag{5.35}$$

Thus, the isospectral deformation offers a method by which we can construct a family of ES potentials which are strictly isospectral to the given potential $V^-(x)$. This has been used to construct conditionally ES rational potentials with EOPs as solutions [42].

5.7.2 Isospectral Shift Deformation

The isospectral deformation of a shape-invariant potential does not lead to shape invariant potentials. But, we can construct another family of ES potentials, which is shape invariant, by modifying the condition in (5.26) to

$$\tilde{V}^+(x) = V^+(x) + R \tag{5.36}$$

where R is a constant. This method is called the isospectral shift deformation method. Substituting $\tilde{V}^+(x)$ and $V^+(x)$ respectively from (5.5) and (5.27) in (5.36) gives

$$\phi^2(x) + 2\,W(x)\phi(x) + \phi'(x) = R \tag{5.37}$$

This again is a Riccati equation, solving which we can determine the unknown function $\phi(x)$, which will be different from that given in (5.30). The details of this method are given in the later sections. The spectra of $V^-(x)$ and $\tilde{V}^-(x)$ have a relative shift of a constant R.

Since this method demands that $\tilde{V}^+(x)$ be shifted by a nonzero constant and not be exactly equal to $V^+(x)$, it is being called the isospectral shift deformation (ISD) method [19].

5.7.3 ISD and Shape Invariant Extensions

Consider the QHJ equation (5.17) and its superpotential solution $\omega(x, \lambda)$. We construct a most general solution $\bar{w}(x, \lambda)$ of the form

$$\bar{w}(x, \lambda) = w(x, \lambda) + Q(x, \lambda). \tag{5.38}$$

demanding that the potential $\bar{V}(x)$ associated with (5.17) be SI. This implies

$$\bar{w}^2(x, \tau(\lambda)) + \bar{w}'(x, \tau(\lambda)) = \tilde{w}^2(x, \lambda) - \tilde{w}'(x, \lambda) + K. \tag{5.39}$$

Here K is independent of x. Substituting (5.38) in (5.39) gives

$$Q(x, \lambda)^2 + 2\omega_0(x, \lambda)Q(x, \lambda) + Q'(x, \lambda) = Q(x, \tau(\lambda))^2 + 2\omega_0(x, \tau(\lambda))Q(x, \tau(\lambda))$$
$$+ Q'(x, \tau(\lambda)) + K. \tag{5.40}$$

Here we have used the fact that if the extended potential is to be SI, then

$$Q(x, \lambda) = -Q(x, \tau(\lambda)). \tag{5.41}$$

The form of (5.40) suggests that it is sufficient to look for a solution such that both sides of (5.41) are equal to a constant, say R_1. Thus we obtain

$$Q(x, \lambda)^2 + 2\omega_0(x, \lambda)Q(x, \lambda) + Q'(x, \lambda) = R_1. \tag{5.42}$$

Comparing the above equation with (5.28), we can identify $Q(x, \lambda)$ with $\phi(x)$. Thus, by demanding that the extended potential be shape invariant, we are led to the condition (5.36), imposed to construct an extended superpotential in the ISD method. Therefore,

$$\tilde{V}_2(x, \tau(\lambda)) = V_2(x, \tau(\lambda)) + R_1. \tag{5.43}$$

Thus we see that if a given potential $V^-(x)$ is shape invariant, we can construct an extended potential $\tilde{V}^-(x)$ which is also shape invariant. In the next section, we explicitly construct the rational extensions of the radial oscillator.

5.8   Radial Oscillator and Its Shape Invariant Extensions

We begin with the supersymmetric radial oscillator potential

$$V^-(r) = \frac{1}{4}\omega^2 r^2 + \frac{\ell(\ell+1)}{r^2} - \omega\left(\ell + \frac{3}{2}\right) \tag{5.44}$$

and its superpotential

$$W(r) = \frac{\omega r}{2} - \frac{l+1}{r}. \tag{5.45}$$

Its shape invariant partner, $V^+(r)$, is given in (5.15). SUSY between $V^\pm(r)$ is exact with respect to the above superpotential.

Construction of the superpotentials
The QMF corresponding to the bound states of the radial oscillator as written in Chap. 3, Sect. 5.2 is

$$q(r) = -W(r) + \frac{P_n'(r)}{P_n(r)}, \tag{5.46}$$

Table 5.1 superpotentials, potentials $V^-(r)$.

i	$W_i(r)$	$V_i^-(r) = W_i^2(r) - \partial_r W_i(r)$
1	$\frac{1}{2}\omega r - \frac{(\ell+1)}{r}$	$V(r) - \omega(\ell + 3/2)$
2	$\frac{1}{2}\omega r + \frac{\ell}{r}$	$V(r) + \omega(\ell - 1/2)$
3	$-\frac{1}{2}\omega r - \frac{(\ell+1)}{r}$	$V(r) + \omega(\ell + 3/2)$
4	$-\frac{1}{2}\omega r + \frac{\ell}{r}$	$V(r) - \omega(\ell - 1/2)$

where $W(r)$ is the superpotential given in (5.45). The polynomials $P_n(r)$ coincides with the associated Laguerre polynomials.

From our earlier discussion on radial oscillator, the general form of $W(r)$ is

$$W(r) = -\frac{b_1}{r} + d_1 r \qquad (5.47)$$

with constants b_1 and d_1 having the following dual values

$$b_1 = -(\ell + 1), \ell \; ; \; d_1 = +\omega, -\omega \qquad (5.48)$$

The four superpotentials for the radial oscillator are listed in the Table 5.1. The third column in the table has the potentials $V_i^-(r)$ corresponding to $W_i(r)$ in the second column. It may be noted that all $V_i^-(r)$ are equal to $V(r)$ apart from a constant. Among the superpotentials listed in Table 5.1, $W_1(r)$ and $W_4(r)$ lead to normalizable ground states of $V_1^-(r)$ and $V_4^+(r)$ respectively. The superpotentials $W_2(r)$ and $W_3(r)$ lead non-normalizable ground states for both the partners and hence correspond to broken SUSY.

We consider $W_2(r)$ and explicitly construct the extended superpotential $\tilde{W}_2(r)$ using ISD method. Identifying $W_2(r)$ with $W(x)$ of (5.25), we set

$$\tilde{W}_2(r) = W_2(r) + \phi(r), \qquad (5.49)$$

where $\phi(r)$ is fixed by demanding

$$\tilde{V}_2^+(r) = V_2^+(r) + R. \qquad (5.50)$$

The above condition ensures that the wave functions of $\tilde{V}_2^+(r)$ and $V_2^+(r)$ are identical apart from a normalization constant and with eigenvalues shifted by a constant R. Substituting $\phi(r) = \frac{d}{dr} \log L(r)$, Eq. (5.28) in this case becomes

$$L''(r) + \left(\omega r + \frac{2\ell}{r}\right)L'(r) - RL(r) = 0. \qquad (5.51)$$

Next a change of variable $\eta = -\frac{1}{2}\omega r^2$, reduces it to

$$\eta L''(\eta) + \left(-\eta + \ell + \frac{1}{2}\right)L'(\eta) + \frac{R}{2\omega}L(\eta) = 0. \tag{5.52}$$

Since we are interested only in obtaining rational extensions, we demand that the solution $L(\eta)$ be a polynomial of degree m, which fixes $R = 2m\omega$. The above differential equation for $L(\eta)$ coincides with that of the Laguerre differential equation and therefore we obtain $L(\eta)$ as the associated Laguerre polynomials

$$L(\eta) = L_m^\alpha(\eta), \qquad \alpha = (\ell - 1/2). \tag{5.53}$$

Thus $\phi(r)$ becomes

$$\phi(r) = -\omega r \frac{\partial_y L_m^\alpha(\eta)}{L_m^\alpha(\eta)}. \tag{5.54}$$

The extended superpotential will now be

$$\widetilde{W}_{2,m}(y) = W_2(y) + \omega r \frac{\partial_y L_m^\alpha(y)}{L_m^\alpha(y)}. \tag{5.55}$$

where $y = \frac{1}{2}\omega r^2 = -\eta$. By construction $\widetilde{V}_2^+(r) \equiv V_2^+(r) + R$ and the desired rational extension of the radial oscillator is its partner potential $\widetilde{V}_2^-(r)$. It is obtained by substituting $\widetilde{W}_{2,m}(y)$ in

$$\widetilde{V}_m^-(r) = \widetilde{W}_{2,m}^2(y) - \widetilde{W}_{2,m}'(y) \tag{5.56}$$

as

$$\widetilde{V}_m^{(-)}(r) = V^{(-)}(r) + 2\omega r\left[\omega r + \frac{2\ell}{r}\right]\frac{\partial_y L_m^\alpha(y)}{L_m^\alpha(y)}$$
$$-\omega r \partial_y\left[\omega r \frac{\partial_y L_m^\alpha(y)}{L_m^\alpha(y)}\right] + \left(\omega r \frac{\partial_y L_m^\alpha(y)}{L_m^\alpha(y)}\right)^2. \tag{5.57}$$

The subscript m in $V_m^-(r)$ is introduced to show the dependency on the index m. For $m = 0$, $L_0^\alpha(y) = 1$ and $V_m^-(r)$ reduces to the radial oscillator potential $V^-(r)$. For $m > 0$, we get a family of potentials, $\widetilde{V}_m^{(-)}(r)$, which are rational extensions of the radial oscillator indexed over m.

Wave functions of $\widetilde{V}_m^{(-)}(r)$
The wave function of the rational extension can be calculated using the wave function of the partner $\widetilde{V}_m^{(+)}(r)$. Using the fact that

$$\widetilde{V}^{(+)}(r) = \frac{1}{4}\omega^2 r^2 + \frac{\ell(\ell-1)}{r^2} + 2m\omega = V^{(+)}(r) + R, \tag{5.58}$$

by construction, eigenfunctions of $\widetilde{V}^{(+)}(r)$ are same as $V^{(+)}(r)$ as given below

$$\widetilde{\psi}_n^{(+)}(r) \equiv \psi_n^{(+)}(r) = y^{\ell/2}\exp(-y/2)L_n^\alpha(y)|_{y=\frac{1}{2}\omega r^2} \qquad (5.59)$$

apart from a normalization constant. These are obtained by replacing $\ell \to \ell - 1$ in the eigenfunctions of the radial oscillator derived in Chap. 2.

The eigenfunctions of $\widetilde{V}_m^{(-)}(r)$ can be calculated by using the SUSY intertwining relation between the solutions of the partners $\widetilde{V}^{(\pm)}(r)$. Thus using

$$\widetilde{\psi}_{n,m}^{(-)}(r) = \left(-\frac{d}{dr} + \widetilde{W}_{2,m}(r)\right)\widetilde{\psi}_n^{(+)}(r), \qquad (5.60)$$

and substituting (5.59) in (5.60) and making use of the recurrence relations of the Laguerre polynomials [26] gives the eigenfunctions of the rational potential (5.57) as

$$\widetilde{\psi}_{n,m}^{(-)}(r) = \left[\frac{r^{(\ell+1)/2}\exp(-\frac{1}{4}\omega r^2)}{L_m^\alpha(-y)}\left(L_m^{(\alpha+1)}(-y)L_n^\alpha(y) - L_m^\alpha(-y)\partial_y L_n^\alpha(y)\right)\right]_{y=\frac{1}{2}\omega r^2}, \qquad (5.61)$$

which can be written as

$$\widetilde{\psi}_{n,m}^{(-)}(r) = \left(\frac{r^{\ell/2}\exp\left(-\frac{1}{4}\omega r^2\right)}{L_m^\alpha(-y)|_{y=\frac{1}{2}\omega r^2}}\right)\widetilde{P}_{n,m}(r). \qquad (5.62)$$

Here $\widetilde{P}_{n,m}(r)$ are obtained by replacing y by $\frac{1}{2}\omega r^2$ in $\widetilde{P}_{n,m}(y)$, where

$$\widetilde{P}_{n,m}(y) = \left[L_m^{(\ell+\frac{1}{2})}(-y)L_n^{(\ell-\frac{1}{2})}(y) - L_m^{(\ell-\frac{1}{2})}(-y)\partial_y L_n^{(\ell-\frac{1}{2})}(y)\right]. \qquad (5.63)$$

These are known as the $L1$-type exceptional Laguerre polynomials. To know more about the orthogonality properties, recurrence relations, Rodriguez formula, etc., we refer the reader to [8].

5.9 Further Observations

In this section, we give some observations related to the process of constructing rational extensions *i.e.*, of going from $V^-(r)$ to $\widetilde{V}^-(r)$ outlined for the radial oscillator explicitly.

The first observation is the existence of more than one type of the generalized family of rational potentials and the corresponding EOPs. In the previous section, the extension of the radial oscillator resulted in a family of rational potentials with $L1$-type Laguerre EOPs as solutions. Similar rational extensions involving $L2$ and

$L3$ type EOPs can be constructed using the superpotentials $W_2(r)$ and $W_3(r)$ listed in Table 5.1. For more details on the construction of these other families of extended potentials and their solutions using ISD, we refer the reader to [19]. Similarly rational extensions of all ES supersymmetric potentials have been constructed and will involve solutions in terms of exceptional Jacobi and exceptional Hermite polynomials. Within the QHJ formalism no new inputs are required and the extension process will be same as that done in the case of the radial oscillator potential.

The second observation is the existence of nontrivial supersymmetry between $\tilde{V}^\pm(r)$. The crucial link in this process is the supersymmetric partner of $V^-(r)$. The SUSY between the two potentials $V^\pm(r)$ is defined in terms of the intertwinning operators which require the superpotential as the input. And the obvious choice of the superpotential is the log derivative of the ground state wave function. There is another suspersymmetry between $V^+(\equiv \tilde{V}^+(r))$ and $\tilde{V}^-(r)$. The corresponding intertwinning operators require the use of superpotential $\tilde{W}_2(r)$ which is not related to the ground state of V^+ in an obvious manner. Hence this is referred to as nontrivial SUSY [43].

The third observation is regarding the family of potentials indexed over a continuous parameter. The generalized family of rational extensions constructed in Sect. 5.8 was indexed over the discrete index m. We can also generate families of extended potentials indexed by a continuous parameter by removing the constraint that the solution of Eq. (5.52) be a polynomial. It is observed that in this case, ensuring that these potentials are nonsingular will require some additional work [44].

The final observation is about the existence of multi-indexed EOPs. From the construction of rational extensions discussed, it is very clear that these can be constructed in a simple and straight forward manner using the ISD taking crucial inputs from the QHJ analysis of the potential models. We can continue this process and construct rational extensions to all these ES rational potentials by applying ISD iteratively and constructing rational potentials which will have multi-indexed EOPs as solutions. This has been explicitly done in [45] and by other methods [40–47].

References

1. Gómez-Ullate, D., Kamran, N., Milson, R.: An extended class of orthogonal polynomials defined by a Sturm-Liouville problem. J. Math. Anal. Appl. **359**, 352–367 (2009)
2. Quesne,C.: Exceptional orthogonal polynomials, exactly solvable potentials and supersymmetry. J. Phys. A: Math. Theor. **41**, 392001–392007 (2008). https://doi.org/10.1088/1751-8113/41/39/392001; Quesne,C,: Solvable rational potentials and exceptional orthogonal polynomials in supersymmetric quantum mechanics. SIGMA **5**, 084 (2009)
3. Infeld, L., Hull, T.E.: The factorization method. Rev. Mod. Phys. **23**, 21–68 (1951)
4. Cooper, F., Khare, A., Sukhatme, U.P.: Supersymmetry and quantum mechanics. Phys. Rep. **251**, 267–385 (1995)
5. Cooper, F., Khare, A., Sukhatme, U.P.: Supersymmetric Quantum Mechanics. World Scientific Publishing Co. Ltd. Singapore (2001)
6. Sree Ranjani, S., et al.: The exceptional orthogonal polynomials, QHJ formalism and the SWKB quantization condition. J. Phys. A: Math Theor. **45**, 055210 (2012)

7. Sasaki, R., Tsujimoto, S., and Zhedanov, A.: Exceptional Laguerre and Jacobi Polynomials and the corresponding potentials through Darboux Crum transformations. J. Phys. A: Math. Theor. **43**, 315204 (20pp) (2010)

8. Odake, S., Sasaki, R.: Infinitely many shape invariant potentials and new orthogonal polynomials. Phys. Lett. B **679**, 414–417 (2009)

9. Gomez-Ullate, D., Kamran, N., Milson, R.: Exceptional orthogonal polynomials and the Darboux transformation. J. Phys. A **43**, 434016 (2010)

10. Grandati, Y.: New rational extensions of solvable potentials with finite bound state spectrum. Phys. Lett. A **376**, 2866–2872 (2012)

11. Grandati, Y.: Solvable rational extensions of the isotonic oscillator. Ann. Phys. **326**, 2074–2090 (2011)

12. Bagchi, B., Grandati, Y., Quesne, C.: Rational extensions of the trigonometric Darboux-Pöschl-Teller potential based on para-Jacobi polynomials. J. Math. Phys. **56**, 062103 (11pp) (2015)

13. Ho, C.-L.: Prepotential approach to solvable rational extensions of Harmonic Oscillator and Morse potentials. J. math. Phys. **52**, 122107 (2011)

14. Grandati, Y.: Rational extensions of solvable potentials and exceptional orthogonal polynomials. J. Phys. Conf. Ser. **343**, 012041 (12pp) (2012)

15. Gómez-Ullate, D., Kamran, N., Milson, R.: Two-step Darboux transformations and exceptional Laguerre polynomials. J. of Math. An. App. **387**, 410–418 (2012)

16. Gómez-Ullate, D., Grandati, Y., Milson, R.: Rational extensions of the quantum harmonic oscillator and exceptional Hermite polynomials. J. Phys. A: Math. Theor. **47**, 015203 (2014)

17. Marquette, I., Quesne, C.: Two-step rational extensions of the harmonic oscillator: exceptional orthogonal polynomials and ladder operators. J. Phys. A: Math. Theor. **46**, 155201 (2013)

18. Quesne, C.: Rationally-extended radial oscillators and Laguerre exceptional orthogonal polynomials in kth-order SUSYQM. Int. J. of Mod. Phys. **26**, 5337–5347 (2011)

19. Sree Ranjani, S., Sandhya R., Kapoor,A.K.: Shape invariant rational extensions and potentials related to exceptional polynomials. Int. Jour. Mod. Phys. A **30**, 1550146 (22pp) (2015)

20. Gendenshtein, L.E.: Derivation of exact spectra of the Schrodinger equation by means of supersymmetry. JETP Lett. **38**, 356–359 (1983)

21. Gelfand, Y.A., Likhtman, E.P.: Extension of the algebra of Poincare group generators and violation of p invariance. JETP Lett. **13**, 323–326 (1971)

22. Wess, J., Zumino,B.: Supergauge transformations in four-dimensions. Nucl. Phys. B **70**, 39–50 (1974); ibid.: Supergauge Invariant Extension of Quantum Electrodynamics. B **78**, 1–13 (1974)

23. Witten, E.: Dynamical breaking of supersymmetry. Nucl. Phys. B **188**, 513–554 (1981)

24. Sandhya, R., Sree Ranjani, S., Kapoor, A.K.: Shape invariant potentials in higher dimensions. Ann. Phys. **359**, 125–135 (2015)

25. Dass, T., Sharma, S.K.: Mathematical Methods in Classical and Quantum Physics. Universities Press (India) Limited, Hyderabad (1998). (This book covers several important results in this area)

26. Dennery, P.A., Kryzwicki.: Mathematics for Physicists. Dover Publications, New York (1967)

27. Bochner, S.: Uber Sturm-Liouvillesche Polynomsysteme. Math. Z. **29**, 730–736 (1929)

28. Routh, E.J.: On some properties of certain solutions of a differential equation of the second order. London Math. Soc. **16**, 245–261 (1885)

29. Odake, S., Sasaki, R.: Another set of infinitely many exceptional X_l Laguerre polynomials. Phys. Lett. B **684**, 173–176 (2010)

30. Odake, S., Sasaki, R.: Infinitely many shape-invariant potentials and cubic identities of the Laguerre and Jacobi polynomials. J. Math. Phys. **51**, 053513 (2010)

31. Ho, C-L., Odake, S., Sasaki, R.: Properties of the exceptional X_l Laguerre and Jacobi polynomials. SIGMA **7**, 107 (24pp) (2011)

32. Gómez-Ullate, D., Kasman, A., Kuijlaars, A.B.J., Milson, R.: Recurrence relations for exceptional Hermite polynomials. J. App. Th. **204**, 1–16 (2016)

33. Ho, C.-L., Sasaki, R.: Zeros of the exceptional Laguerre and Jacobi polynomials. ISRN Math. Phys. **2012**, Article ID 920475 (2012)

34. Dutta, D.: On the completeness of exceptional orthogonal polynomials in quantum systems. Int. J. App. Math. **26**, 601–609 (2013)
35. Durán, A.J.: Exceptional Charlier and Hermite orthogonal polynomials. J. App. Th. **182**, 29–58 (2014)
36. Liaw, C., Littlejohn, L.L., Milson, R., Stewart, J.: The spectral analysis of three families of exceptional Laguerre polynomials. J. App. Th. **202**, 5–41 (2016)
37. Durán, A.J.: Exceptional Hahn and Jacobi polynomials with an arbitrary number of continuous parameters. J. App. Th. **214**, 9 (45pp) (2017)
38. Durán, A.J.: Integral Transforms and Special Functions, vol. 26 (2015)
39. García-Ferrero, M.A. Gomez-Ullate, Milson, D.R.: Exceptional Gegenbauer polynomials via isospectral deformation. arXiv:2110.04059v1
40. Sasaki, R.: Universe **2**, 2 (2014)
41. Pappademos, J., Sukhatme, U.P., Pagnamenta, A.: Bound states in the continuum from super-symmetric quantum mechanics. Phys. Rev. A **48**, 3525 (1993)
42. Dutta, D., Roy, P.: Conditionally exactly solvable potentials and exceptional orthogonal poly-nomials. J. Phys. A: Math. Theor. **51**, 042101 (2010)
43. Shiv Chaitanya, K.V.S., Sree Ranjani, S., Panigrahi, P.K., Radhakrishnan, R., Srinivasan, V.: Exceptional polynomials and SUSY quantum mechanics. Pramana J. Phys. **85**, 53–63 (2015)
44. Odake, S., Sasaki, R.: A new family of shape invariantly deformed Darboux Pöschl Teller Potentials with Continuous ℓ. J. Phys. A **44**, 195203 (14pp) (2011)
45. Sree Ranjani, S.: Quantum Hamilton-Jacobi Route to Exceptional Laguerre polynomials and the corresponding rational potentials. Pramana-J. Phys. **93**, 29 (14pp) (2019)
46. Odake, S., Sasaki, R.: Exactly solvable quantum mechanics and infinite families of multi-indexed orthogonal polynomials. Phys. Lett. B **702**, 164–170 (2011)
47. Odake, S.: Recurrence relations of the multi-indexed orthogonal polynomials. III. J. Math. Phys. **57**, 023514 (2016)

Chapter 6
Complex Potentials and Optical Systems

6.1 Introduction

In this chapter we first illustrate the efficacy of QHJ formalism in dealing with complex potentials involving parity and time reversal ($\mathcal{PT}$ symmetries). The $\mathcal{PT}$-symmetric complex potentials have been the subject of intense investigation in recent times due to the appearance of both real and complex eigenvalues [1]. The QHJ approach, with its natural use of the complex variables, is found to be quite useful in throwing light on the deeper structure of these quantum problems.

6.2 Complex $\mathcal{PT}$-Symmetric Potentials

We start with the $\mathcal{PT}$-symmetric potentials [2–4], having both real and complex spectra, originating from the nontrivial pole structure of these complex potentials. The nature of the QMF and its comparison with the case of real potentials is highlighted, for a deeper understanding of the underlying analytic structure.

The QHJ equation reads

$$p^2 + \frac{\hbar}{i}p' = 2m(E - V) \tag{6.1}$$

Clearly the left-hand side of this equation is invariant under the transformations $x \to -x$ and $i \to -i$, provided the quantum momentum function behaves analogously. This implies that if the potential on the right-hand side is $\mathcal{PT}$-symmetric then the QHJ equation will remain invariant for real energy spectrum. We will now look at an explicit example of a $\mathcal{PT}$-symmetric potential.

© The Author(s), under exclusive license to Springer Nature Switzerland AG 2022
A. K. Kapoor et al., *Quantum Hamilton-Jacobi Formalism*,
SpringerBriefs in Physics, https://doi.org/10.1007/978-3-031-10624-8_6

6.2.1   The Complex Scarf-II Potential

In this section we examine the potential known as the complex Scarf-II potential given by [4]

$$V(x) = A \operatorname{sech}^2(x) + i B \operatorname{sech}(x) \tanh(x) \tag{6.2}$$

Evidently it is invariant under the transformations $x \to -x$ and $i \to -i$ and hence is a $\mathcal{PT}$-symmetric potential. Defining $q = \dfrac{d \ln \psi}{dx}$ gives the QHJ equation

$$q^2 + q' + E - V(x) = 0, \tag{6.3}$$

where $V(x)$ is defined in Eq. (6.2). With a change of variable $y = i \sinh x$ along with

$$\chi = \left(\phi - \frac{y}{2\left(1 - y^2\right)} \right) \tag{6.4}$$

$$q = i \left(\sqrt{1 - y^2} \right) \phi. \tag{6.5}$$

Equation (6.3) transforms as

$$\chi^2 + \frac{d\chi}{dy} + \frac{2 + y^2}{4\left(1 - y^2\right)^2} - \frac{E}{1 - y^2} - \frac{A - By}{\left(1 - y^2\right)^2} = 0. \tag{6.6}$$

It is easy to see that χ in Eq. (6.4) has poles at $y = \pm 1$ apart from the n moving poles with residue one. Under the assumption that these poles are the only singularities, we can write down χ as a sum of its singular and analytical parts as

$$\chi = \frac{a_1}{y - 1} + \frac{b_1}{y + 1} + \frac{P_n'}{P_n} + C, \tag{6.7}$$

The residues at $y = \pm 1$ are easily computed and we get

$$a_1 = \frac{1}{2} \pm \frac{1}{2}\sqrt{\frac{1}{4} + A - B} \tag{6.8}$$

and

$$b_1 = \frac{1}{2} \pm \frac{1}{2}\sqrt{\frac{1}{4} + A + B}. \tag{6.9}$$

Note that a_1 and b_1 contain two tunable parameters A and B. These will be important in the subsequent discussion.

Using a similar technique as in earlier chapters, the residue at infinity is computed. Equating the sum of all residues to zero we get

$$-E = \left(a_1 + b_1 + n - \frac{1}{2}\right)^2. \tag{6.10}$$

and the wave function in terms of a_1, b_1 is

$$\psi(y) = \exp\left(\int dy \left(\frac{a_1}{y-1} + \frac{b_1}{y+1} + \frac{P'_n}{P_n} + \frac{y}{2(1-y^2)}\right)\right)$$
$$= (y-1)^{a_1 - \frac{1}{4}}(y+1)^{b_1 - \frac{1}{4}} P_n(y). \tag{6.11}$$

To evaluate the unknown polynomial P_n the value of χ from Eq. (6.7) is substituted into Eq. (6.6). Substituting the expression for E from Eq. (6.10) into the differential equation for P_n transforms it in the form of a Jacobi differential equation. Thus the polynomial P_n is Jacobi polynomial $P_n(y) = P_n^{2a_1 - 1, 2b_1 - 1}(y)$ and the wave function is found to be

$$\psi(x) = (i\sinh x - 1)^{b_1 - \frac{1}{4}}(i\sinh x + 1)^{b'_1 - \frac{1}{4}} P_n^{2b_1 - 1, 2b'_1 - 1}(i\sinh x). \tag{6.12}$$

6.2.2 Variation of Parameters and Energy Spectrum

Recall that the residues a_1 and b_1 contain two adjustable parameters A and B. The correct value for each residue is found using the fact that the wave function must vanish at infinity, that is, $\psi(x) \rightarrow 0$ as real x tends to ± 1. We consider two cases of the potential parameters A and B (Fig. 6.1).

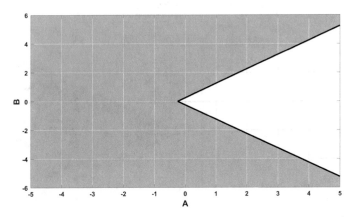

Fig. 6.1 Shaded region represents the $A - B$ parameter space for which $|B| > A + \frac{1}{4}$. We get a complex energy spectrum in this case

Case 1: $|B| > A + \frac{1}{4}$

In this case the possible values of the residues at $y = \pm 1$ are $a_1 = \frac{1}{2} \pm \frac{i}{2}$ $\sqrt{B - A - \frac{1}{4}}$ and $b_1 = \frac{1}{2} \pm \frac{1}{2}\sqrt{A + B + \frac{1}{4}}$. The square integrability of ψ rules out one of the roots of b_1 leaving us with

$$a_1 = \frac{1}{2} \pm \frac{i}{2}\sqrt{B - A - \frac{1}{4}} \tag{6.13}$$

$$b_1 = \frac{1}{2} - \frac{1}{2}\sqrt{A + B + \frac{1}{4}}, \tag{6.14}$$

subject to $2n < s - 1$ with

$$r = \sqrt{B - A - \frac{1}{4}}, \quad s = \sqrt{A + B + \frac{1}{4}}, \tag{6.15}$$

the wave function and energy spectrum are given by

$$\psi = (i \sinh x - 1)^{\frac{1}{4} \pm \frac{ir}{2}} (i \sinh x + 1)^{\frac{1}{4} - \frac{s}{2}} P_n^{\pm ir, s}(i \sinh x), \tag{6.16}$$

$$E = \left(n + \frac{1}{2} - \frac{1}{2}(s \pm ir)\right)^2, \tag{6.17}$$

Note the complex nature of the energy spectrum in this case.

Case 2: $|B| \leq A + \frac{1}{4}$ (Fig. 6.2).

In a manner similar to the previous case, we can rule out one of the values from a_1 leaving us with

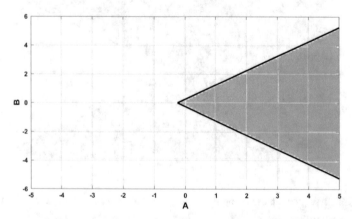

Fig. 6.2 Shaded region represents the $A - B$ parameter space for which $|B| \leq A + \frac{1}{4}$. We get a real energy spectrum in this case

$$a_1 = \frac{1}{2} - \frac{1}{2}\sqrt{\frac{1}{4} + A - B},$$ (6.18)

$$b_1 = \frac{1}{2} - \frac{1}{2}\sqrt{\frac{1}{4} + A + B},$$ (6.19)

such that the wave function and energy turn out to be

$$\psi = (i\sinh x - 1)^{\frac{1}{2} - \frac{\mu}{2}} (i\sinh x + 1)^{\frac{1}{2} - \frac{\nu}{2}} P_n^{\mu,\nu}(i\sinh x)$$ (6.20)

and

$$E = -\left(n + \frac{1}{2} - \frac{1}{2}(\mu + \nu)\right)^2$$ (6.21)

with

$$\mu = \sqrt{\frac{1}{4} + A - B} \text{ and } \nu = \sqrt{\frac{1}{4} + A + B}.$$ (6.22)

For these values of parameters, the energy spectrum is real with n restricted by

$$2n < \mu + \nu - 1.$$ (6.23)

The fact that calculations in QHJ formalism inherently take into account the complex plane and that QHJ equation admits $\mathcal{PT}$- symmetry allowed us to study the complex Scarf-II potential, which is a $\mathcal{PT}$-symmetric potential with energy in complex as well as real sectors based on the value of tunable parameters of the potential.

6.3 SUSY, $\mathcal{PT}$-Symmetry and Optical Systems

In this section we will investigate $\mathcal{PT}$-symmetry in the context of optical systems where the Helmholtz equation becomes analogous to the Schrödinger equation with the refractive index profile playing the role of a quantum mechanical potential. As an explicit example we consider the Scarf-II potential which exhibits supersymmetry with both broken and unbroken phases [5].

Interestingly, supersymmetry has been successfully used for controlled lasing and other optical applications. It has been found very useful in phase-matched mode selection, removal of undesired modes and other phase-sensitive applications. It plays a crucial role in designing $\mathcal{PT}$ inspired optical couplers [6–8]. Both these symmetries are characterized by unbroken and broken phases, having different physical significance which can be implemented in optical systems.

As an example, Miri et al. [9] have shown that SUSY-generated isospectral complex refractive index profiles can be constructed for removal of the fundamental or a higher-order guided mode. Consider an eigenvalue problem of the form

$$H\psi = \left(-\frac{d^2}{dX^2} - V(X)\right)\psi = -\lambda\psi, \tag{6.24}$$

where $X = x/x_0$ is the coordinate transverse to the wave guide axis, scaled arbitrarily by x_0 and which supports at least one guided mode $\psi_1^{(1)}(X)$ with eigenvalue $\lambda_1^{(1)}$. The Hamiltonian $H^{(1)} = H + \lambda_1^{(1)}$ can be factorized as $A_2 A_1$ with

$$A_1 = \frac{d}{dX} + W, \qquad A_2 = -\frac{d}{dX} + W, \tag{6.25}$$

where in general, $A_2 \neq A_1^\dagger$ for a complex superpotential W. The partner Hamiltonian $H^{(2)} + \lambda_1^{(1)} = A_1 A_2$ can be used to generate the isospectral optical potentials (apart from fundamental mode of the original potential) through

$$V^{(1,2)} = \lambda_1^{(1)} - W^2 \pm W'. \tag{6.26}$$

An example of $\mathcal{PT}$-symmetric multimode wave guide with four guided modes, having a modulated higher-order Gaussian refractive index profile, is

$$\Delta n^{(1)}(x) = \delta\left(1 + i\gamma\tanh\frac{k_0 x}{2\pi w_I}\right)\exp\left[-\left(\frac{k_0 x}{2\pi w_R}\right)^8\right]. \tag{6.27}$$

It is appropriately parameterized by the index elevation δ, gain-loss contrast γ, geometrical parameters w_R and w_I. The free-space wave number k_0 of its superpartner wave guide has only three guided modes as the partner spectra are related by

$$\lambda_m^{(1)} = \lambda_{m-1}^{(2)}, \quad m > 1. \tag{6.28}$$

Further, Miri et al. [9] show that using a higher order mode to construct the SUSY-partner gives the freedom to remove the corresponding higher-order mode in the spectrum of the partner.

We start by noting that the mapping of the Helmholtz equation in optical systems to the Schrödinger's equation allows us to use the tools developed in quantum mechanics to study optical systems. We consider a monochromatic electromagnetic plane wave incident in the $x - y$, with the wave vector k_0 making an angle θ with the x- axis, in an inhomogeneous dielectric medium with a distribution

$$\epsilon(x) = \epsilon_b + \alpha(x), \tag{6.29}$$

where ϵ_b is the background (substrate's) dielectric constant and the inhomogeneity in the dielectric distribution is $\alpha(x)$. The inhomogeneity disappears at $|x| \to \infty$ giving $\epsilon(x) \to \epsilon_b$ and the tangential component of electric field, $\mathcal{E}_z$, exhibits continuity. We can write the Helmholtz equation for $\mathcal{E}_z$ as

$$\frac{d^2\mathcal{E}_z}{dx^2} + \left(k_0^2\epsilon(x) - k_y^2\right)\mathcal{E}_z = 0, \tag{6.30}$$

where $k_y = k_0\sqrt{\epsilon_b}\sin\theta$ is the tangential component of the wave vector. The term $k_0^2\epsilon_b - k_0^2\epsilon(x) \equiv V(x)$ is the analog of the quantum mechanical potential $V(x)$ and the refractive index $(n(x) = \sqrt{\epsilon(x)\mu(x)})$ is given by

$$n^2(x) = n_b^2 - \frac{V(x)}{k_0^2}. \tag{6.31}$$

Here $n_b = \sqrt{\epsilon_b}$ is the refractive index of the background and for dielectric materials $\mu(x) = 1$.

The dielectric distributions of the form in Eq. (6.29) were studied by Love and Ghatak [10] for exact solutions of transverse magnetic (TM) modes in graded index wave guides under the constraint

$$\frac{d^2\epsilon(x)}{dx^2} - \frac{3}{2}\frac{1}{\epsilon}\left(\frac{d\epsilon(x)}{dx}\right)^2 - 2p\epsilon = 0, \tag{6.32}$$

where p is a constant. Equation (6.32) has two solutions

$$\epsilon(x) = \epsilon(0)\exp\left(-2|x|/a\right) \tag{6.33}$$
$$\epsilon(x) = \epsilon(0)\operatorname{sech}^2\left(x/a\right). \tag{6.34}$$

The potential $V(x)$ with $\epsilon(x)$ as given in equation (6.34) has been studied in the context of reflectionless potentials for designing anti-reflection coatings across a wide range of wavelengths [11, 12]. Recently, Laba and Tkachuk [13] explored the analogy of transverse electric (TE) and TM modes in planar wave guides having arbitrary spatial variation of permittivity and permeability with the time-independent Schrödinger equation. They concluded that the TE and TM modes in such waveguides are related by a SUSY transformation subject to the spatial variation of permittivity and permeability leading to a spatially homogeneous refractive index. Chen et al. have shown that bright solitons are supported in $\mathcal{PT}$-symmetric generalized Scarf-II potentials in one- and two-dimensional configurations with focusing Kerr-type non-linearity [14]. Non-Hermitian (lossy) optical systems integrated with those having optical gain exhibit exceptional point singularities when tuned to balance the loss and gain. Such designs are used to create metamaterial structures with novel optical properties [15] and are analogous to $\mathcal{PT}$-symmetry breaking in $\mathcal{PT}$-symmetric open quantum systems [16] and can form the basis of sensing applications. Spectral phase transitions also provide the ground for ultrafast sensing applications in optical systems like optical parametric oscillators as well as SUSY-QM systems [3, 17]. The requirement from $\mathcal{PT}$-symmetry is that $n(x) = n^*(-x)$ which sets a constraint on the type of potential that can be used. For explicitness, we now study the refractive index profile of the form given by the familiar Scarf-II potential to explore and relate the broken and unbroken SUSY and $\mathcal{PT}$-symmetry.

6.4 Complex Scarf-II Potential: $\mathcal{PT}$ Symmetry and Supersymmetry

6.4.1 Broken and Unbroken Phases of $\mathcal{PT}$-Symmetry

In order to study the broken and unbroken sectors of $\mathcal{PT}$ [3] and SUSY [5], we introduce a parameter $C^{\mathcal{PT}}$ in the superpotential which now reads

$$W_{\mathcal{PT}}^{\pm} = \left(A \pm iC^{\mathcal{PT}}\right) \tanh \alpha x + \left(\pm C^{\mathcal{PT}} + iB\right) \operatorname{sech} \alpha x. \tag{6.35}$$

Here all the parameters are real. The potential V_- corresponding to the above superpotential is

$$\begin{aligned} V_-(x) = &-\left[\left(A \pm iC^{\mathcal{PT}}\right)\left(A \pm iC^{\mathcal{PT}} + \alpha\right) - \left(\pm C^{\mathcal{PT}} + iB\right)^2\right] \operatorname{sech}^2 \alpha x \\ &-i\left(\pm iC^{\mathcal{PT}} - B\right)\left[2\left(A \pm iC^{\mathcal{PT}}\right) + \alpha\right] \operatorname{sech} \alpha x \tanh \alpha x \\ &+ \left(A \pm iC^{\mathcal{PT}}\right)^2. \end{aligned} \tag{6.36}$$

This potential has both real and complex spectra. For the potential to be $\mathcal{PT}$-symmetric, the coefficient of the first term must be real and the second imaginary, $V_-(x)$. These conditions lead to

$$C^{\mathcal{PT}}[2(A - B) + \alpha] = 0. \tag{6.37}$$

$C^{\mathcal{PT}} = 0$ corresponds to unbroken $\mathcal{PT}$-symmetry leading to bound states with a real spectrum. However, SUSY is broken in some parametric domains. $C^{\mathcal{PT}} \neq 0$ corresponds to the spontaneous $\mathcal{PT}$-broken phase, with complex eigenvalue spectrum. Therefore, $C^{\mathcal{PT}}$ behaves like an order parameter, with its nonvanishing value characterizing broken $\mathcal{PT}$ phase.

Interestingly, in both, unbroken and broken phases of $\mathcal{PT}$-symmetry, the potential $V_-(x)$ can be obtained from two distinct superpotentials, due to its underlying symmetry. These two superpotentials generate two disjoint sectors of the Hilbert space in the unbroken case. Here, the spectrum is real and shape invariance leads to translational shift in a suitable parameter by real units. For a different parametrization, two different superpotentials lead to the same potential for the broken $\mathcal{PT}$-symmetry as well. In this case, SI produces complex parametric shifts. We present the interconnection between unbroken $\mathcal{PT}$-symmetric and broken SUSY as well as spontaneously broken $\mathcal{PT}$ phases in the following sections, and for more details see [3].

6.4.2 Unbroken $\mathcal{PT}$ and Broken SUSY Phase

For $C^{\mathcal{PT}} = 0$, we get a $\mathcal{PT}$- symmetric potential

$$
\begin{aligned}
V_{\mathcal{PT}}(x) = A^2 &- \left[A(A + \alpha) + B^2\right] \text{sech}^2 \alpha x \\
&+ i B(2A + \alpha) \, \text{sech} \, \alpha x \tanh \alpha x
\end{aligned}
\tag{6.38}
$$

derived from the superpotential

$$
W_1(x) = A \tanh \alpha x + i B \, \text{sech} \, \alpha x. \tag{6.39}
$$

We notice that there is a choice of transformation that keeps the superpotential invariant. The transformation $A \leftrightarrow B - (\alpha/2)$, provides us with another candidate as the superpotential

$$
W_2(x) = \left(B - \frac{\alpha}{2}\right) \tanh \alpha x + i \left(A + \frac{\alpha}{2}\right) \text{sech} \, \alpha x. \tag{6.40}
$$

Both, $W_1(x)$ and $W_2(x)$ give us the same $\mathcal{PT}$−symmetric potential and lead to two sets of real energy eigenvalues

$$
E_n^1 = -(A - n\alpha)^2, \tag{6.41}
$$

$$
E_n^2 = -\left(B - \frac{\alpha}{2} - n\alpha\right)^2, \tag{6.42}
$$

Provided $A > 0$ and $B > \alpha/2$ (white region in Fig. 6.3), respectively. For the remaining region in the A-B parameter space (grey region in Fig. 6.3), at least one of the energy eigenvalue set fails to give a normalizable ground state, indicating broken SUSY.

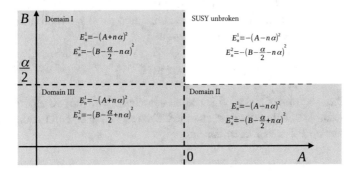

Fig. 6.3 Parameter space with $C^{\mathcal{PT}} = 0$. The white and grey regions correspond to the unbroken and broken SUSY phases

Table 6.1 Shape invariance and energy eigenvalues in the parameter domains of broken SUSY phase.

Domain	Region	Shape invariance	E_n^1	E_n^2
I	$A < 0$, $B > \alpha/2$	† $A \to -A$ $A \to A + \alpha$	$-(A + n\alpha)^2$	unchanged
II	$A > 0$, $B < \alpha/2$	† $B - \alpha/2 \to -(B - \alpha/2)$, $A + \alpha/2 \to -(A - \alpha/2)$; $B - \alpha/2 \to B - \alpha/2 + \alpha$, $A - \alpha/2 \to A - \alpha/2$	unchanged	$-\left(B - \frac{\alpha}{2} + n\alpha\right)^2$
III	$A < 0$, $B < \alpha/2$	Both techniques, one for Domain I for the eigenvalues of $W_1(x)$ and one for Domain II for the eigenvalues of $W_2(x)$	$-(A + n\alpha)^2$	$-\left(B - \frac{\alpha}{2} + n\alpha\right)^2$

† To flip the sign and take it to the unbroken SUSY domain

The exact energy eigenvalues for the broken SUSY cases can be obtained by two SI transformations as given in Table 6.1. For convenience, the energy eigenvalues in the various parameter domains have been diagrammatically shown in Fig. 6.3.

6.4.3 Broken $\mathcal{PT}$-Symmetric Phase

For the order parameter $C^{\mathcal{PT}} \neq 0$, that is, the spontaneous broken $\mathcal{PT}$ phase case, we get, $A = B - (\alpha/2)$. Using this in Eq. (6.35) gives

$$W_{\mathcal{PT}}^{\pm} = \left(A \pm iC^{\mathcal{PT}}\right)\tanh \alpha x + \left(\pm C^{\mathcal{PT}} + i\left(A + \frac{\alpha}{2}\right)\right)\operatorname{sech}\alpha x. \qquad (6.43)$$

Both the superpotentials lead to same $\mathcal{PT}$-symmetric complex potential corresponding to broken $\mathcal{PT}$- symmetry. Imposing the condition for shape invariance leads to the complex energy eigenvalues in the spectra

$$E_n^{\pm} = -\left(A - n\alpha \pm iC^{\mathcal{PT}}\right)^2. \qquad (6.44)$$

Note that the spectrum of energy eigenvalues associated with the two superpotentials $W_{\mathcal{PT}}^{\pm}$ are complex conjugates of each other. The respective eigenfunctions, related to each other by $\mathcal{PT}$ transformations, are

$$\psi_n^{\pm}(x) \propto (\operatorname{sech}\alpha x)^{(1/\alpha)\left(A \pm iC^{\mathcal{PT}}\right)}$$
$$\times \exp\left[-\frac{i}{\alpha}\left(A + \frac{\alpha}{2} \mp iC^{\mathcal{PT}}\right)\tan^{-1}(\sinh \alpha x)\right]$$
$$\times P_n^{\mp(2iC^{\mathcal{PT}}/\alpha),-(2A/\alpha)-1}(i\sinh \alpha x). \qquad (6.45)$$

In the broken $\mathcal{PT}$ domain, supersymmetry remains intact. It is interesting to note that complex eigenmodes with real energy eigenvalues arise when $A = n\alpha$ in this domain, corresponding to zero-width resonances [18].

All the solutions obtained above for the various phases of $\mathcal{PT}$ symmetry and supersymmetry can be obtained using the QHJ formalism. This potential is a good theoretical contender for modeling in optical systems.

References

1. Bender, C.M., Boettcher, S.: Real spectra in non-Hermitian Hamiltonians having $\mathcal{PT}$ symmetry. Phys. Rev. Lett. **80**, 5243 (1998)
2. Ahmed, Z.: Real and complex discrete eigenvalues in an exactly solvable one-dimensional complex PT-invariant potential. Phys. Lett. A **282**, 343 (2001)
3. Abhinav, K., Panigrahi, P.K.: Supersymmetry, PT-symmetry and spectral bifurcation. Ann. Phys. **325**, 1198 (2010)
4. Sree Ranjani, S., Kapoor, A.K., Panigrahi, P.K.: Quantum Hamilton-Jacobi analysis of PT symmetric Hamiltonians. Int. J. Mod. Phys. A **20**, 4067 (2005)
5. Bhalla, R.S., et al.: Quantum Hamilton–Jacobi analysis of phases of supersymmetry in quantum mechanics. Int. J. Mod. Phys. A **12**, 1875 (1997)
6. Lupu, A., Benisty, H., Degiron, A.: Using optical PT-symmetry for switching applications. Photonics Nanostruct. Fundam. Appl. **12**(4), 305 (2014)
7. El-Ganainy, R., Makris, K.G., Christodoulides, D.N., Musslimani, Z.H.: Theory of coupled optical PT-symmetric structures. Opt. Lett. **32**, 2632 (2007)
8. Principe, M., Castaldi, G., Consales, M., Cusano, A., Galdi, V.: Supersymmetry-inspired non-Hermitian optical couplers. Sci. Rep. **5**, 1 (2015)
9. Miri, M.A., Heinrich, M., Christodoulides, D.N.: Supersymmetry-generated complex optical potentials with real spectra. Phys. Rev. A **87**, 043819 (2013)
10. Love, J., Ghatak, A.: Exact solutions for TM modes in graded index slab waveguides. IEEE J. Quantum Electron **15**, 14 (1979)
11. Gupta, S.D., Agarwal, G.S.: A new approach for broad-band omni directional antireflection coatings. Opt. Express **15**, 9614 (2007)
12. Gupta, S.D. et al.: Wave Optics: Basic Concepts and Contemporary Trends. CRC Press (2015)
13. Laba, H.P., Tkachuk, V.M.: Quantum-mechanical analogy and supersymmetry of electromagnetic wave modes in planar wave guides. Phys. Rev. A **89**, 033826 (2014)
14. Chen, Y., et al.: Soliton formation and stability under the interplay between parity-time-symmetric generalized scarf-II Potentials and Kerr non linearity. Phys. Rev. E **102**, 012216 (2020)
15. Miri, M.A., Alu, A.: Exceptional points in optics and photonics. Science **363**, eaar7709 (2019)
16. Eleuch, H., Rotter, I.: Open quantum systems with loss and gain. Int. J. Theor. Phys **54**, 3877 (2015)
17. Roy, A., et al.: Spectral phase transitions in optical parametric oscillators. Nat. Commun. **12**, 1 (2021)
18. Mostafazadeh, A.: Optical spectral singularities as threshold resonances. Phys. Rev. A **83**, 045801 (2011)

Chapter 7
Beyond One Dimension

7.1 Introduction

In the previous chapters, solutions to a variety of quantum models have been presented within the QHJ formalism. The methods discussed so far can also be applied to separable systems in higher dimensions in a straightforward manner. The formalism has been very successful in obtaining solutions for eigenvalues and eigenfunctions of a wide variety of one-dimensional potential models. It is also found useful in the construction and study of rational extensions of potential models related to the exceptional orthogonal polynomials discovered and actively investigated in the past few years.

Ushveridze [1] has investigated and obtained extensive results on the classification of ES and QES models in one and higher dimensions. In the next section we briefly summarize the main idea of Ushveridze's work. Finally, some open problems are listed in the concluding section Sect. 7.3.

7.2 Classification of ES and QES Models

A detailed investigation of classification of ES and QES models has been carried out by Ushveridze. The details of analytic and algebraic approaches to this problem and the important results are given in Chaps. 3 and 4 of the monograph by Ushveridze [1]. See also references cited in these chapters. We will briefly sketch the analytic method described in Sect 3.5 onwards.

The analytic approach and the QHJ formalism have two important points in common. The first one is the QHJ equation itself, and the second one being use of the analytic properties of the QMF in the complex plane.

The study of the ES and the QES models by Ushveridze aims at the classification of the models and does not attempt to formulate an exact quantization condition, or to

A. K. Kapoor et al., *Quantum Hamilton-Jacobi Formalism*,
SpringerBriefs in Physics, https://doi.org/10.1007/978-3-031-10624-8_7

give a scheme of computation of eigenvalues and eigenfunctions within the analytic method.

To understand the essential points of the analytic approach, it is first to be noted that the Schrodinger equation for a potential $V(x)$, in one dimension,

$$\left(-\frac{d^2}{dx^2} + V(x) - E\right)\phi(x) = 0,$$
(7.1)

is a special case of a more general equation

$$\left[-\frac{d^2}{dx^2} + \omega(x)\right]\phi(x) = 0,$$
(7.2)

where $\omega(x)$ is a function of the form

$$\omega(x) = \epsilon_0 \chi^0(x) - \sum_{\alpha=1}^{D} \epsilon_\alpha \chi^\alpha(x).$$
(7.3)

where the functions $\chi^\alpha(x), \alpha = 0, 1, \ldots, D$ are some potential like functions that do not depend on the solution $\phi(x)$. The parameters ϵ_α are spectral parameters and have different values for different solutions $\phi(x)$. An equation such as Eq. (2.31) can arise from separation of variables in higher dimensions. Solutions of the Eq. (2.46) are examined in the complex plane by introducing the logarithmic derivative

$$y(x) = \frac{d\log \phi(x)}{dx} = \frac{1}{\phi(x)}\frac{d\phi(x)}{dx}.$$
(7.4)

keeping the notation of Ushveridze. It is apparent that $y(x)$ is just the same as QMF for multiparameter spectral equation 2.31, still in one complex variable. Equation (2.46), written in terms of $y(x)$ take the form

$$y^2(x) + y'(x) = \omega(x)$$
(7.5)

That this is just the QHJ equation cannot be missed. For lack of any name used in [1], we continue to call this equation as the QHJ equation. The fixed singularities of $y(x)$ come from the fixed singularities of $\omega(x)$. Besides the fixed singular points, the function $y(x)$ will have moving poles with unit residues. Thus the function $y(x)$ has the form

$$y(x) = F(x) + \sum_{k}^{M} \frac{1}{x - \xi_k}, \quad M = 0, 1, \ldots.$$
(7.6)

where $F(x)$ is the fixed singular part of $y(x)$ and the sum is over the moving poles. The location of the moving poles, in general, depends on the solution $\phi(x)$. A straight forward analysis of constraints imposed by QHJ equation gives the following three relations. The first one is about the solution $\phi(x)$

$$\phi(x) = \exp\left(\int F(x)\,dx\right) \Pi_{k=1}^{M}(x - \xi_k).$$ (7.7)

The second relation is a set of equations giving the constraints on the location of the moving singularities ξ_k

$$F(\xi_j) + \sum_{k \neq j, k=1}^{M} \frac{1}{\xi_j - \xi_k} = 0, \quad j = 1, \ldots, M.$$ (7.8)

The third relation expresses the function $\omega(x)$ in terms of the function $F(x)$

$$\omega(x) = F^2(x) + \frac{dF(x)}{dx} + 2\sum_{k=1}^{M} \frac{F(x) - F(\xi_k)}{x - \xi_k}.$$ (7.9)

The analytic method as presented in [1] does not start with a general form for the potential (or of functions $\chi^\alpha(x)$), nor does it attempt to derive constraints on the form of the potential. The heart of the classification is the function F corresponding to the fixed singular part of $y(x)$. This function F is assumed to be a rational function. Everything else is determined once F is known; Eq. (2.20) gives the solution for the wave function $\phi(x)$ and 2.13 give expressions for the functions $\chi^\alpha(x)$ in terms of the function F. Equations 2.12 constrain the locations ξ_k, $k = 1, \ldots, M$ of the moving poles. In the following paragraph, a set of requirements is described which constrain the form of $F(x)$.

The function $F(x)$ is assumed to be a *meromorphic function* of a complex variable x, and consequences of Eqs. (2.18), (2.12) and (2.13) are worked out. It is assumed that $F(x)$ is a rational function of x and is, therefore, completely specified in terms of finite number of unknown parameters. The unknown parameters in $F(x)$ are (i) the order of pole at infinity, (ii) the location of the poles of $F(x)$ and (iii) the corresponding residues. Ushveridze looks for solutions for functions $\chi^{(\alpha)}$, $\alpha = 1, \ldots, D$ subject to the set of restrictions given above. Equation 2.13 shows that $\chi^\alpha(x)$, in Eq. (2.31), must also be rational functions. In addition, it is required that the solution for these functions must not depend on the location of moving poles ξ_k. The location of the moving poles is constrained to satisfy Eq. (2.12).

A graphical analysis of all restrictions and a physical requirement that the solution $\phi(x)$ may be element of some Hilbert space are investigated. A lengthy analysis leads to a classification of possible forms of the function $F(x)$.

It will be seen that the classification of the function $F(x)$ solves the problem of classification of the ES and QES problems through Eqs. (2.18), (2.20) and (2.13). While Eq. (2.12) restricts the possible location of moving poles ξ_k, Eq. (2.20) gives an explicit expression of solution $\phi(x)$.

These answers, through Eqs. (2.20) and (2.13), give a solution to the classification problem in one dimension. This classification leads to the potential $V(x) \equiv \chi^{(0)}(x)$ and of functions $\chi^{(\alpha)}(x), \alpha = 1, \ldots, D$ as rational functions of x. In order to see the familiar examples of ES potentials coming out of this work, a change of variable is required. More details of this can be found in Sect. 3.5 to Sect. 3.8. Section 3.9 of [1] gives results for higher dimensional models . Details about classification of QES models are given in Chap. 4 [1]. Full details of the analysis and a complete set of results are available only in the original articles listed in the bibliography.

7.3   Open Questions and Concluding Remarks

We have seen that QHJ formalism provides an elegant and simple framework for investigation of eigenvalues and eigenfunctions of one-dimensional models and for classification of ES and QES models. We list some of the challenging problems that remain.

1. The QMF for the bound states has a simple, meromorphic form. It will be inter-esting to investigate the singularity properties of QMF for scattering states and to ask if any simplification occurs for the exactly solvable scattering problems in one dimension.
2. For the periodic potential models, the eigenfunctions could be easily obtained for the band edges. Within the QHJ formalism, it is still an open problem to investigate the singularity structure of QMF and to solve for the wave functions for energy values inside the allowed bands.
3. At present nothing is known about generalizing the QHJ scheme to higher dimen-sions. In particular, it is of interest to ask if one could formulate a scheme of computing energy eigenvalues without first solving for the eigenfunctions. Such a study will require answers to several questions. For example,

   • Should the properties of the solutions be studied in several complex variables? Or in some, suitably chosen one complex variable?
   • Does a simple and tractable form, like meromorphic function in one dimension, of QMF emerge in higher dimensions?
   • How does one formulate and implement quantization and boundary conditions for the bound states?
   • The QMF in one dimensions is defined as the logarithmic derivative of the quantum action. The corresponding object in n- dimensions will have n- com-ponents, one component corresponding to each partial derivative of the quan-tum action function. In our view a straightforward approach of investigating the properties of multicomponent QMF may not be the best starting point.

At present nothing is known about possible answers to the above questions within the QHJ formalism. It may be useful to note that, for higher-dimensional models, some exact results have been given within the path integral approach. We refer the reader to [2] and to the appendix of Chap. 6 [3]. Coming back to the QHJ formulation, it remains a challenge to even identify the best starting point and to make progress. Finally, a successful investigation of the above questions is likely to lead to construction of such models.

References

1. Ushveridze, Alexander G.: Quasi-Exactly Solvable Models in Quantum Mechanics. Institute of Physics Publishing, Bristol and Philadelphia (1993)
2. Schulman, L.S.: Exact time-dependent green's function for the half-plane barrier. Phys. Rev. Lett. **49**, 599–601 (1982)
3. Schulman, L.S.: Techniques And Applications Of Path Integration. John Wiley and Sons, Inc. (1981)

Index

© The Editor(s) (if applicable) and The Author(s), under exclusive license to Springer 111
Nature Switzerland AG 2022
A. K. Kapoor et al., *Quantum Hamilton-Jacobi Formalism*,
SpringerBriefs in Physics, https://doi.org/10.1007/978-3-031-10624-8

Printed in the United States
by Baker & Taylor Publisher Services